the COCKROACH papers

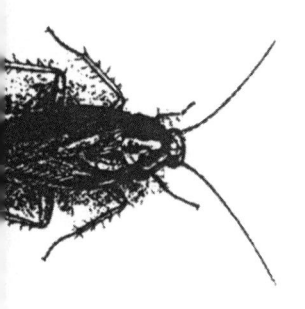

the **cockroach** papers

A COMPENDIUM OF HISTORY AND LORE

with a new Preface
richard **schweid**

The University of Chicago Press
Chicago and London

The University of Chicago Press, Chicago 60637
The University of Chicago Press, Ltd., London
Copyright © 1999 by Richard Schweid
All rights reserved.
University of Chicago edition 2015
Printed in the United States of America

24 23 22 21 20 19 18 17 16 15   1 2 3 4 5

ISBN: 978-0-226-26047-1 (paper)
ISBN: 978-0-226-26050-1 (e-book)
DOI: 10.728/chicago/9780226260501.001.0001

Originally published by Four Walls Eight Windows in 1999.

Illustration credits: p. 7, 120–64 (versos) from *The Cockroach*; pp. 10, 11, 13, 14, 20, 33, 57, 58, 89, 91, 93 © Dr. Betty Faber; p. 65 © Louay Henry; p. 116 © Tony Langford; p. 168 (top) from Backyard Brains product page of the RoboRoach; p. 168 (bottom) © Shannon Marlow du Plessis.

Library of Congress Cataloging-in-Publication Data
Schweid, Richard, 1946– author.
   The cockroach papers : a compendium of history and lore, with a new preface / Richard Schweid.
      pages : illustrations ; cm
   Includes bibliographical references and index.
   ISBN 978-0-226-26047-1 (pbk. : alk. paper) — ISBN 978-0-226-26050-1 (ebook) 1. Cockroaches. 2. Cockroaches—Anecdotes. 3. Human-animal relationships. I. Title.

QL505.5.S38 2015
595.7'28—dc23
                                                                    2015010572

Neapolitan folk saying:

**ogni scarrafone è bello
a mamma sua.**

**every cockroach is beautiful
to its mother.**

bello mio!

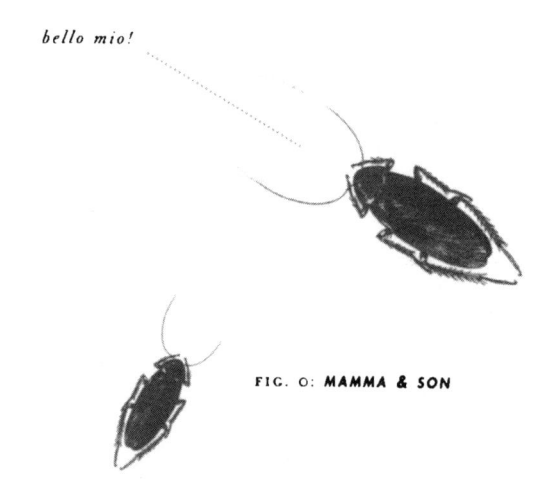

FIG. O: *MAMMA & SON*

acknowledgments

IT IS CUSTOMARY HERE TO THANK PEOPLE for helping me along the bumpy way that is any sort of writing project, and to declare that any mistakes, inaccuracies, or unpleasantries within these pages are wholly my own fault and responsibility. How much more necessary both of these tasks become if the subject of the writing elicits the sorts of reactions common when dealing with cockroaches.

In addition to all those who were patient enough to answer my questions during interviews, thanks for kindness and assistance go out to José Manuel Álvarez Flórez, Montse Ayuso, Carlos Bosch, Luisa and Vincenzo d'Arista, Susan and James Edmunds, Diane and Ezra Eichelberger, Celia Filipetto, James and Terilynn Graham Freedman, Donna Glassford, Mike Golden, Jeff Judd, Sue Katz, Silvia Komet, Alan and Andrée LeQuire, Carmen Martínez, Emanuele Nastasi, Joan Salvat, Jay Sapir, Adele, Jean and David Schweid, Dan Stein, and Ornella Zoia. And, thanks to editor Kathryn Belden for her hard work to make the book better.

Contents

preface

WHEN I WAS WORKING ON THIS BOOK, in 1999, people often asked why I wanted to spend my time writing about something as repulsive as a cockroach.

The first reason was, of course, simply a desire to learn about the natural history of such a ubiquitous insect, but other things also drew me. One was that humans forget the vast majority of moments from their daily lives, whole days, weeks, months go by about which we can remember nothing, but most Westerners will always remember the encounters they've had with cockroaches. The "civilized" European or North American has such a visceral distaste for the cockroach that a meeting with one is likely to be burned into his or her memory bank. This meant that just about everyone had a good, skin-crawling cockroach story to tell.

Another draw to the subject was the long, long road that humans and cockroaches have traveled together. They were here waiting for us when our species arose, and we have never separated from them. If, and when, we disappear from the earth, it is improbable that cockroaches will accompany us into extinction. At least not right away. They are considerably hardier than we are, with a much more viable design for survival.

Yet another reason I liked writing about cockroaches was that what I wrote would stay relevant for a long time. Books often come with a kind of use-by date. Many subjects are ephemeral, they change substantially over the course of a couple of decades. Readers' tastes change, and some subjects lose their attraction. In other cases, discoveries are made that render previous thinking and writing on the subject outmoded. None of this applies to cockroaches. What's true about them today is likely to be true tomorrow, and next year, and next millennium, since they've been around, basically unchanged, for 3.5 million years. That's more than ten times as old as the earliest human remains. Not everything in this book has stayed the same since I wrote it, but all the basic stuff about cockroaches is unchanged. Cockroaches keep their heads down, and can eat just about anything, a couple of great survival strategies. They are not likely to disappear, or even to change much, anytime soon. A minor genetic mutation here or there, perhaps, but the nuts and bolts of their natural history will stay the same.

What *has* changed over the past couple of decades is my life and those places in it where cockroaches once thrived. The locations I wrote about are not the same now as they were then, nor am I, but overall the world is pretty much as it was then. If peace has come to one place, war has erupted in another, and if my bank account is bigger than it was then, my knees are weaker.

A few minor changes about some of the subjects touched on in this book need noting: the Minnesota company H. B. Fuller has stopped selling addictive shoe glue to Latin American markets (street children who are addicted to the toluene in the glue have no trouble finding other brands that will do just as well, and are still, literally, huffing their brains out); the New York City Housing Authority now uses more gels than sprays to combat roaches in public housing; the Environmental Protection Agency tightened its regulations regarding the testing of pesticides on human beings in 2013; and the wonderful cockroach researcher Lou Roth passed away in June 2003. Other things, unfortunately, remain unchanged: young women still die violently in disproportionately high numbers in Ciudad Juárez;

a permanent peace agreement has yet to be signed in the Western Sahara; and the rate of asthma among poor children exposed to cockroach waste continues to rise.

Roach investigation is ongoing in many places: research on controlling the movement of roaches remotely to help locate survivors after earthquakes continues at places like North Carolina State University. A hardy cockroach called *P. japonica*, previously only seen in the Far East, hitched a ride to Manhattan last year and may be settling in near the High Line park. And, scientists continue to use cockroaches as subjects in studying topics as various as the connection between learning and time of day (this at Vanderbilt University) and DNA mapping (e.g., the National Cockroach Project at Rockefeller University, which used citizen scientists to gather data from 2004 to 2012). Reviewing preliminary data from that project, a senior research associate named Mark Stoekle commented, "This is a window into cockroach society, and it is very much like our own."

Some minor changes have also occurred in the relationship of *Homo sapiens* to cockroaches. Continued funding for arms research has produced a steady flow of more sophisticated weapons against roaches, but often when something is found that works initially, it loses effectiveness over the course of a few years, as cockroaches mutate and develop resistance to the new product. The uneasy coexistence between humans and roaches, with its frequent skirmishes, continues unabated, as it has for millennia.

In places where spending money to exterminate roaches is unthinkable because there's barely enough, or sometimes not enough, to pay for food and shelter, coexistence with cockroaches is obligatory. In developed countries, where household budgets are high enough to allow for extermination measures, sprays and powders are often used because they are cheaper than gel baits. However, professional exterminators in developed countries have moved almost exclusively to gels as they are less noxious to humans and more efficient.

An improvement in the weapons we use against cockroaches has not tilted the battle toward us. When a more effective roach killer is developed, the targeted species often mutates remarkably quickly to combat the noxious effects. Over the course of only five years, a strain of German cockroaches in Florida developed an aversion to the glucose that previously had been used to attract them to a poisonous bait. Researchers at Michigan State University have found another strain with a mutated gene giving it a high " knockdown resistance" to pyrethroid insecticides. An article in the October 2013 *Journal of Economic Entomology* reported that still other strains of German cockroaches, collected in the field, were found to have developed resistance to many classes of insecticides, including chlorinated hydrocarbons, organophosphates, carbamates, pyrethroids, phenylpyrazoles, and oxadiazines. The authors concluded: "Insecticide resistance has become a major problem for the pest management industry."

As if this ability to mutate rapidly in response to various insecticides was not enough bad news for fastidious folks, the weapons we use against cockroaches have often proven to be more harmful to us than to the bugs. For instance, recent research has shown that exposure of women in the third trimester of pregnancy to some of the standard ingredients used in pyrethroid insecticides can result in mental developmental delays for a child after birth. Others classes of roach killers are no safer: organophosphates and N-methyl carbamates also have a potential to harm the nervous systems of children. Regardless of the class of insecticide, it is designed to disrupt the roach's neurological system and disable its vital organs. It only makes sense to avoid contact with something capable of doing that to anything as hardy as a cockroach.

Sprays, gels, bug bombs, and poisons have not moved us all that far ahead of the ancient Egyptians who prayed to the ram-headed god Khnum for protection against cockroaches thousands of years ago; or the beleaguered John Smith writing from the Virginia colony in 1624: "A certaine India Bug, called by the Spaniards a caca-roche, the which creeping into Chests they eat and defile with their ill-sented dung."

Human beings are pretty much as besieged by roaches now

as we were then, part of a larger struggle against the insect kingdom, a battle in which we have been engaged since making our appearance on this planet. As if to confirm that any victories we may achieve in the bug wars will prove temporary, the bedbug has made a tremendous comeback in the Western world over the past two decades. This has reversed a steep decline in the bedbug population over the twentieth century, and they have reappeared to colonize the mattresses of the rich and poor alike, infesting five-star hotels and homeless shelters, anywhere people lie down to take their rests. The rise in bedbug numbers is thought to be associated with two factors: more people traveling from one place to another, and increased insecticide resistance.

Bedbugs make roaches look like amiable fellow travelers. While a cockroach is not averse to nibbling human flesh, it does not drink human blood, as does the parasitic bedbug. What's more, the bedbug is much more patient about waiting for its food than a cockroach. A bedbug can hold off feeding for months while it waits for a human being to make his or her warm blood available. But, of course, the terrible itching that bedbugs can cause is nothing compared to the potentially lethal diseases and fevers spread by mosquito bites. Given their threats to our health and tranquility, it is likely that we will continue to wage war with bugs until humans or insects disappear from our planet, and in the long run there's little doubt that we will *not* be the last ones standing.

Despite the fact that we are acculturated to view them as our mortal enemies, cockroaches have a lot to teach us. For one thing, our self-understanding has been greatly enhanced by studying the way roaches are put together. An incalculable number of two-inch long American cockroaches have been dissected in laboratories to elucidate for researchers our own biological functions, and how our nervous systems are wired. The simple and tremendously efficient design of the cockroach has taught scientists a lot. It is hard to imagine a life-form more apparently distinct from our own, but studying cockroaches has revealed many things we can apply to *Homo sapiens*. Berta Scharrer, one of the founders of the science of neuroendocrinology, and a Nobel nominee, did her research with roaches. She wrote: "Here is an animal of frugal habits, tenacious of life, eager to live in the laboratory and very modest in its space requirements."

The lessons cockroaches can teach us are not simply neurological and biological, not only those things that we learn when we take them apart in search of clues about the structure of life. They also keep us humble, serving as constant reminders that we are vulnerable to tiny life-forms, that while we may be able to misuse our intelligence to build ever more terrible ways to kill each other, our power is not unlimited. Far from it, because a small roach can repulse us, a wee bedbug can torture us, and something as insubstantial as a mosquito can kill us.

The cockroach lives in our dark spaces, it constantly reminds us that other worlds exist at the margins of all our lives, worlds that obey drastically different orders from the one we know. Roaches construct a whole other reality, which they weave into our own. It is there under the sink, behind the refrigerator, inside the pipes, or just on the other side of the baseboards. A world governed by things like animal instinct, finely tuned reflexes, and close quarters. Cockroaches represent the awful thoughts and feelings lying just under each of our surfaces, lurking just behind the facades we present to the world. We know that repressing these dark desires, rather than acknowledging them, can have disastrous consequences for human beings. Like these pools of darkness in our own selves, cockroaches cannot be eliminated from our lives simply by ignoring them.

Nor should they be. We owe these constant companions a bit of our conscious attention, and close observation rewards us with an understanding of a fascinating reality, so close to, and yet so far from, our own. Hopefully, this edition of the book will continue to function as a field guide to that reality, and fulfill the primary purpose for which I wrote it: to introduce readers to the amazing world of the cockroach.

Richard Schweid
Barcelona, 2014

saving all
sentient beings

IN THE SUMMER OF 1967, I was twenty-one years old and living in New York City. I had fled from a long childhood and adolescence in Nashville. I slept in the living room of a small, three-room apartment on the second floor of a Christopher Street brick building, half a block west of Sheridan Square. This place had a narrow metal shower built into a corner of the bedroom and the toilet was in a closet off the living room. The woman who paid the rent did so out of a monthly allowance from her family, and she was sleeping with a friend of mine. He moved in and brought a pair of friends with him, one of whom was myself. She was an open and generous soul, so the living arrangements were fine with her.

There was a sofa in the small living room, and the two of us slept out there. I put the cushions on the floor night after night, month after month, and my friend Jeff did his best to make himself comfortable on the sprung springs of the sofa. Every so often, when we ran smack-dab out of cash, we would walk over to the pier on the Hudson River where ships docked with boxes of produce. A large group of men gathered each evening for a shape-up, which meant standing around in a loose cluster inside a hulking, high-ceilinged wooden warehouse, waiting to see who the foreman would choose to give work to that night. The work entailed loading boxes of fruit and vegetables on to trucks, packing them inside tractor trailers to be hauled across the country. Five dollars an hour, cash at the end of the night.

Work was not a high priority. A half dozen of us spent our time hanging out, drinking cheap wine, smoking good weed, playing music, writing long collective stories, painting together, trying to put rhyme or reason to our lives and the world around us; we were determined to save not only our own asses but those of our friends, neighbors, and every sentient being in that order from the tyranny of history repeating itself, history as dull labor, war, and death. I was usually tired enough and sufficiently substance-saturated by the time everyone else had left or gone to bed, and Jeff and I could dismantle the sofa, that I went right to sleep on the cushions, despite the cracks between them and the narrowness of the platform they provided.

I slept in a T-shirt and underwear, pants and shirt tossed on a chair. There was a particular July morning when I left my dreams behind and woke up, and my first thought, even before opening my eyes was, What a strange feeling: the lightest of ticklings all over my body, as if someone were breathing very gently up and down my stretched-out form. Tiny gusts of air barely ruffling the hairs on my arms and legs. Lazily, I opened my eyes. The evening before, while I had been loading fruit, exterminators had come by and fumigated the building. My supine body was a charnel house, a killing field of dead and dying roaches that had come out from behind the walls, from the dark spaces under the refrigerator and the stove, from all their sanctuaries. They were driven out in confusion as their poisoned bodies broke down, and their nervous systems went haywire. They died slowly, on their backs, legs kicking feebly into the air. The spasmodically jerking legs are what I had felt upon awakening. The roaches covered the floor, thousands of them, and they were dying all over me. I leapt up screaming, my shout open throated and horrified, as if the cushions had suddenly become a bed of hot coals.

I spent many months of my life sleeping on the floor of that apartment, walking through the neighborhood, east to Greenwich Village, west to the Hudson River. I spent hours and days sitting on the stoop watching the weird world of the West Village go past. I was convinced that this was my life, and a worthy one at that, a conviction I can barely remember now. I can vaguely recall how it felt to feel that way, so certain then that so much idle time would bear fruit further down the road, but now all the days of those years are reduced to nothing but a black hole with shards of recollections scattered here and there, bits of colored cloth caught on the jagged edges of what passes for my memory. But one thing I remember as if it happened yesterday was how those roaches felt dying all over my body.

. • • ●

In the store the old men gathered, occupying for endless hours the creaking milkcases, speaking slowly and with conviction upon matters of profound inconsequence, eying the dull red bulb of the stove with their watery vision. . . . In the glass cases roaches scuttled, a dry rattling sound as they traversed the candy in broken ranks, scaled the glass with licoriced feet, their segmented bellies yellow and flat.

from THE ORCHARD KEEPER
by Cormac McCarthy

● ● • .

After a while, the shape-up work loading fruit became kind of discouraging. Some nights there wasn't work, and as the fall evenings grew chillier, it was a cold walk over to the pier, so I looked around for something a little steadier. Waiting on tables seemed like a good idea and I started to walk around the neighborhood, asking. The first waiter's job I landed was at a coffee shop that occupied a corner of Sixth Avenue and West Fourth Street. My debut night on the job, the cook told me to go down in the basement and bring up a sack of potatoes. He was a corpulent black man with strong arms, biceps the size of hams, and a round, bowling-ball head shaved bare. "The light's at the top of the stairs," he said, motioning toward the door to the basement.

I flicked on the switch, and the light illuminated a busy traffic of roaches and rats moving rapidly across the floor at the bottom of the stairs. Lots of animals dine on roaches, including cats, lizards, and monkeys, but they seem to be well out of harm's way in the company of rats, at least where other food is being stored. I was halfway down the stairs before my mind registered what my eyes were seeing and my ears were hearing: the scurrying of a healthy population of both rats and roaches. I turned around and tore back upstairs to bear news of the infestation, shouting, "There's a bunch of roaches and rats downstairs," wondering if the restaurant would have to be closed down while the exterminators were called in to eliminate this obviously state-of-emergency threat to public health.

The cook couldn't stop laughing, even after he'd called in all the rest of the restaurant's staff to tell them how I'd come running back up the stairs yelling that the basement was full of roaches and rats. He laughed until he was wiping away tears with his big white apron, asking, "Where you from anyway, boy? Where you from?" Then he sent me back down for the potatoes.

. . ● ●

Out of the corner of his eye he saw Umbrella Man scoop a roach off the bar in a movement surprisingly swift for one so sluggish—and in the same movement jam it between his teeth.

Frankie's hand stopped on the glass: here came
Umbrella Man, the bug's blood streaking down
teeth and chin and the bug itself crushed—feelers
still waving between the teeth—"Man! Wash!
Gimme wash!"—pleading between the clenched
teeth and his smeared face right up to Frankie's.
Frankie turned his head away, shoved the
beer toward Umbrellas and didn't turn his head
back till he heard Umbrellas drain the glass to
the last drop.
"He never done anything like that before,"
Frankie complained to the widow Wieczorek.
"What's gettin' into him?"
"He does it all the time now," Widow
explained with a certain pride; as if she had
taught him such a trick.

from THE MAN WITH THE GOLDEN ARM
by Nelson Algren

•   •   •   .

Algren was practically impeccable. Not only was he the
hardest punching writer in the United States, as his contemporary Ernest Hemingway said after this book was published, but
he was a master at combining humor and human horror, the
urban novel at its best. A graveyard humor born of tenements,
taverns, and neighborhoods, the low-grade, ongoing scuffle to
survive, and unlike so many fighters who grow old, his punch
never slowed down, his sense of timing never dulled. His penulti-
mate book of fiction, a collection of short stories called THE LAST
CAROUSEL, published in 1973, is Algren at the top of his form.

He uncharacteristically got a detail wrong in the above
passage from his most famous novel, originally published in
1949 and winner of the first National Book Award. Cockroach
blood is a pigmentless, clear substance circulating through the

interior of its body, and what usually spurts out of a roach when its hard, outer shell–its exoskeleton–is penetrated or squashed is a cream-colored substance resembling nothing so much as pus or smegma. Not the dark liquid implied in Algren's evocative description of a repulsive way to cadge a beer, a sequence that, unsurprisingly, was entirely left out of Otto Preminger's watered-down film of the novel, which starred a young Frank Sinatra.

The off-white stuff is actually fat, which encases a cockroach's organs, circulatory and nervous systems, a thick layer of goo between the tough cuticle of its shell and its delicate insides. This fat body, as it is called, is where much of the insect's metabolism goes on, and where it stores precious nitrogens and other nutrients to have on hand in case food gets scarce. In fact, if they have access to water, German cockroaches, *Blattella germanica*, the most common domestic roach in the United States and the species we usually see in our kitchens, have been observed to live forty-five days without food, and with neither food nor water they can still survive more than two weeks. Other species, most notably the *Periplaneta americana*, the second most common domestic roach in the U.S., can live much longer. With water, *Periplaneta* has been observed to make it as long as ninety days without food, and has gone some some forty days in the laboratory with neither food nor water. In all species, the females are able to do without for longer than the males.

The cockroach, regardless of species, is built for survival. This is the case for many insects, but cockroaches, as far as we know, are the oldest insect still abroad on the planet, a tremendously successful design in evolutionary terms. Like all insects, they have six legs and a shell made of a hard substance called chitin. Their heads are permanently bent down beneath their carapaces, or shells, with a pair of antennae sticking out in front. Seen in profile, a cockroach's head is always bowed. Its waxy exoskeleton and its shape allow it to squeeze into extremely small spaces, and it can utilize a tremendous range of substances for nourishment. In the wild, different species of cockroaches eat a

Figure 1. Cockroach fossils.

wide variety of things from plant debris to fungus to wood to animal dung, depending on what is available. While numerous animals are classified as omnivores–meaning they will eat anything–few live up to the name so well as roaches. The handful of pest species that hang around people will eat almost everything a human being will, except for cucumbers, which they are reported to dislike avidly, and they will also gladly eat a large number of things we would not willingly consume even if starving, including glue, feces, hair, decayed leaves, paper, leather, banana skins, other cockroaches, dead or live human beings, and warm sour beer, which is one of their favorites.

There are fossils of cockroaches from the Carboniferous period, dating back to around 325 million B.C. They predated dinosaurs by more than 150 million years, and humans by more than 300 million. Whereas every other insect fossil from that epoch shows an animal that is now extinct, the cockroaches found buried deep in the earth of the Lower Illinois coal measure are little changed from those found today in houses on top of that same ground. They were plentiful during the Carboniferous period, so much so that it is occasionally called the Age of Cockroaches, and they are still plentiful today. More than 5,000 species of cockroach have been discovered and named during the last couple of centuries, and scientists believe that about the same number remain to be found. All cockroaches belong to the order Blattaria, taken from the Greek word *blattae,* which is what the ancient Greeks called the bugs that were their domestic pests.

The closest insect relatives to roaches are termites and the mantids, such as the preying mantis. They, along with crickets and grasshoppers, all share with the roach a mouth that rips, tears, and grinds, but has no teeth, along with broad forewings that are not of much use—often none at all—for flying. All these insects were once grouped under the taxonomic order, Orthoptera, although that has been discarded and the cockroaches given their own. Within Blattaria, there are five families, and of the estimated 10,000 *Blattarian* species in those families, there are only a handful around the world, far less than a hundred, that live near enough to people to ever be seen by them. As might be expected with an animal that loves heat and humidity, tropical jungles are, perhaps, their favorite homes. In 1983, a scientist set six traps out in the jungles of Panama and recovered 164 different species.

However, roaches are not confined to any particular environment and live in a tremendous variety of places, from underneath woodpiles in Alaska to high in the jungle canopy in the tropics of Costa Rica, inside water-filled *Bromeliads* in the rain forests of Trinidad, in underground chambers in Australia, over 7,000 feet up in the mountains of central Asia, in the swamps of Formosa, the caves of Borneo, and under thorn bushes in arid stretches of Kenya. Most of these thousands of different species will never cross paths with a human being. Wherever they live, they are eminently successful at surviving. If there is a God that made all life forms, a particularly rich blessing was bestowed on the roach, because it got the best design of all. It is, undeniably, one of the pinnacles of evolution on this planet. As such, we certainly have more to learn from them than they do from us, and, to prove it, humans have spent a tremendous amount of both time and money studying them. There is a surprisingly vast scientific literature about every imaginable aspect of cockroach biology and behavior. Their reproductive biology was being studied and debated throughout the nineteenth century. Despite their obvious differences from us, they are considered to be excellent models for neurobiology. A roach is, so the thinking goes, pure instinct—they are straightforward, sentient machines, eminently knowable at the biological level. The American cockroach, *Periplaneta americana*, has long been a favorite animal for biology students to take apart because of its substantial size and abundance. "It is a reasonable assumption that more cockroaches have been dissected on the laboratory bench than any other insect and more cockroach mouthparts, too, have been examined and drawn under the microscope than those of any other insect," wrote P.B. Cornwell, in his 1968 book, THE COCKROACH.

There are, plain and simple, a lot of people with Ph.D.s who make good money working day in and day out with cockroaches. Up until the 1970s, most of the research was in areas related to their biology, how their bodies worked, how they reproduced, and what taxonomic species they belonged to. Since then, there has been a sharp increase in the number of studies related to the social and living habits of the cockroach. A good deal of this has been paid for by chemical companies that manufacture and sell insecticides designed to interrupt those very habits, to control and/or eliminate roaches. Since bedbugs virtually disappeared from the developed world over the past fifty years or so, cockroaches have become the most annoying household pest on many people's lists.

So, in addition to scientists, there is another large class of professionals who make a good living from roaches:

exterminators. Pest control, and more particularly doing battle against cockroaches, generates a lot of money every year. Estimates from the United States Department of Agriculture state that some $4 billion a year is spent trying to control roaches, ants, rats, and termites; and a recent study put the amount of that spent trying to exterminate cockroaches at $240 million (Stix, 1994).

The scientific names bestowed on the common pest cockroaches by Swedish naturalist Carolus Linnaeus in the eighteenth century do not accurately reflect their points of origin. For instance, that roach conjured up by Algren on the widow Wieczorek's Chicago bar would have been, most probably, a German cockroach, *Blattella germanica*, now thought to have come with the Phoenicians across the Mediterranean from Africa, and to have spread throughout Russia, Europe, and on to the Americas. This is the one most North American city dwellers see scurrying for cover when they come in at night and turn on the light, and it is the one that provides exterminators with most

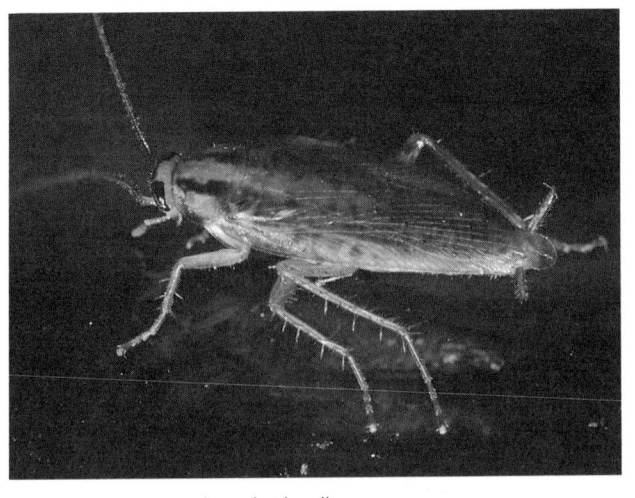

Figure 2. German cockroach *Blattella germanica*. Photo © Betty Faber.

of their work. It is the relatively small, brown cockroach that inhabits apartments and houses, usually the kitchens and bathrooms therein.

The German is one of five species of cockroach that are domestic pests in the United States (there are another sixty-four species living far from populated areas, which most people will never see) and it is the most common. It has wings, but they are vestigial, relics from an earlier form. The German cockroach does not fly. Most species of cockroach have wings, but many of them are strictly for show. The second most common cockroach in the U.S., however, still uses its wings for an occasional flight. It is the above-mentioned *Periplaneta americana*, often euphemistically referred to as a *water bug*, but generally known as the American cockroach, although it is not originally from the Americas, but came from Africa on slave ships. American roaches are dark brown and considerably larger than the German. They grow up to nearly two inches and are likely to be found in southern climes and the subtropics, although they are hardy

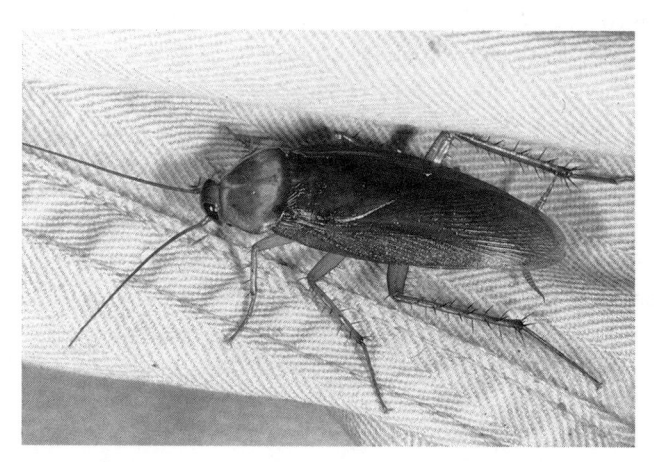

Figure 3. American cockroach "Waterbug" *Periplaneta Americana*.
© Betty Faber.

enough that patient observation of the area around the boiler in the basement of any New York City apartment building is likely to be rewarded with a sighting of an American cockroach. They prefer life in sewers, basements, and other dark, dank places.

I made their acquaintance a few years back, when I was working a six-month stint as a wage-slave copy editor in Brownsville, Texas, and that is when I learned that these sumbucks don't scurry when you turn on the light, they fly, although they can't keep it up for long and latch on to the first thing they encounter. Many a night coming back beaten down and worn out after sending the next morning's edition of the BROWNSVILLE HERALD to press, brain burned by hours in front of the cold light of the computer screen, I would arrive home wanting nothing so much as a sit-down with a cold beer. I would unlock the front door, flip the light switch, and something that seemed the size of a small bird would whir past my head. Or a number of somethings, one after the other. As many times as it happened, nearly each of the 180 nights I lived there, I never got accustomed to the sight and sound of those things, wings outstretched as the long brown bodies, startled by the burst of light, launched themselves through the air to land where they would—on a counter, in my hair, clinging vertically to the wall above the sofa, wherever. The best thing about them was that they died easily. The barest tap with a rolled newspaper, even a paper as thin and lightweight as the HERALD, was all it took to render one dead.

The third species occasionally encountered by North Americans is the oriental cockroach (Blatta orientalis), which generally lives in outdoor garbage and is the most common household species in Europe. The fourth is the brown-banded roach (Supella longipalpa), which is not so widespread, but seems to prefer living in libraries or dens. It is also much given to colonizing appliances, which provide it with a much-appreciated warmth. Around a refrigerator motor, or inside television sets, are likely places to have an infestation of them. A fifth species, the smokey-brown (Periplaneta fuliginosa), is usually

Figure 4. Brown-banded cockroach *Supella longipalpa*. © Betty Faber.

found south of the Mason-Dixon line, often spends the warm months outside, and when the weather begins to get chilly makes its way indoors.

. . • •

Regardless of what the species is, they all have fat bodies. They act as reserves for the roach—in times of plenty, lipids and proteins are stored there, and if food disappears, these can be called on to fuel muscular activity and keep the body nourished. Its function is somewhat analogous to our livers, and the fat body is, in fact, considered to be an organ unto itself. A part of the digestive process—the conversion of food into amino acids—takes place there, as do a number of other metabolic processes, including the breakdown of toxic materials.

Most people who see a cockroach in their home have an almost instinctive urge to step on, squash, swat, or otherwise kill it, but there is no denying that the object of so much revulsion is a marvelously designed envelope for its DNA. The fat body may

Figure 5. Roach nymph holding an aggressive stance. © Betty Faber.

look like nothing more than creamy, thick jism, but it is a key to the design that has made the cockroach one of the world's most successful life-forms, here long before *Homo sapiens* arrived and with every prospect of still being here when humans are long gone.

Among the nutrients stored in the fat body against a time of scarcity are uric acid salts. Instead of expelling uric acid from the body with its feces, a roach stores a substantial amount of it in the fat body, and these reserves are called urates. There have been a variety of hypotheses as to why a roach does this, the most recent explanation being that prehistoric cockroaches survived on a scavenged, low-nitrogen diet and that when they encountered something to eat with a higher-than-normal nitrogen content, they were able to store it in the form of urates in the fat body to be used later, when nitrogen-rich nurtrients were not available. Studies have shown that when a roach is consuming a diet high in nitrogen, urate levels increase, and decrease when the diet is low in nitrogen (Ross and Mullins, 1995).

Another use for the urates appears to be connected to reproduction (Mullins and Keil, 1980). While there are species of cockroaches that give birth to live young, the vast majority reproduce by way of an egg case, called an ootheca, which contains the fertilized eggs and protects them as they develop into embryos. In some species, the female keeps this case in her body until just before the nymphs inside it are ready to come forth into the world, and in other species, the egg case is deposited in what the roach perceives to be a safe and concealed place where it will remain until it hatches. There is substantial evidence that urates are passed to the female by the male during copulation, as part of the sperm packet, or separately depending on the species, and play a role in the formation of the egg case.

Other constituents of the fat body contribute to a roach's metabolic process, such as proteins, lipids, and carbohydrates. These are stored when not needed, and in times of scarcity they are released to provide energy. All of these have been finely measured in the laboratory, by starving roaches for various lengths of time and analyzing their fat bodies at various stages. The numerous metabolic reactions that take place there produce substances that enter the circulatory system.

Nelson Algren did not get it entirely wrong—cockroaches do have blood, or at least a circulatory fluid, which constantly bathes the fat body. This hemolymph, as it is called, is clear in color. What gives our blood its color is the oxygen-carrying hemoglobin in it. Hemolymph does not carry oxygen to the tissues like our own blood does. In fact, oxygen and respiration in the cockroach are wholly different from the system of heart and lungs as we know it.

A cockroach has no lungs and does its breathing through its body. There are ten pairs of openings, called *lateral spiracles,* along the edges of the cockroach's abdomen, each pair more or less opposite another, with two pairs up front at the thorax and the other eight spaced out along, approximately, the back half of the insect. The entrance of air into the spiracles, which stay closed when not in use to prevent the loss of body moisture, is accomplished in one of two ways: if a roach is at rest with a low oxygen requirement, it can do perfectly well on the amount of air that casually enters the spiracles when they are open. If, however, it is spending a lot of energy, or the temperature is over about 85 degrees Fahrenheit, it will need an additional quantity of oxygen. Contraction and relaxation of abdominal muscles provide the mechanical ventilation needed to draw air into the spiracles. Regardless of how air enters the spiracles, once it is there it passes through the thin walls of the tubes, which branch out like dendrites or coral in a tracheal network through the body, and washes over the surrounding tissues. During this process, carbon dioxide and water are produced. Cockroaches can remain alive "for a number of hours" without any oxygen, according to Cornwell, who notes that if they are kept in an atmosphere of entirely nitrogen, when they are released back into the air they consume an excess of oxygen equal to that of which they have been deprived.

While cockroach blood does not carry oxygen, the hemolymph does serve other familiar sanguinary tasks, such as transporting nutrients and waste products through the roach's body, sealing wounds by coagulating at the point of damage, and carrying hormones from the glands that secrete them to the receptors where they need to go. A cockroach has a tube-shaped heart that pumps the hemolymph toward the head with each of its contractions, forcing a flow throughout the body. Blood enters the twelve-chambered heart through twelve pairs of openings that act as control valves, preventing the hemolymph from flowing back through them and forcing it out of the heart on its one-way circuit of the roach's body.

.   •   •   •

Water is a precious thing to a cockroach, so precious, in fact, that if a roach does not have an ample amount of it around, it will not excrete liquids. Rectal pads, located almost at the end of the animal's excretory system, squeeze water from the mass to be excreted just before elimination. This liquid gets recycled to places like the fat body, and the insect's only excretion will consist of dry solids. Cockroaches do not waste water, and if given a choice they will always make their homes near some: in drains, in a stored sack of slowly softening potatoes, or in a cabinet under a sink where condensation and tiny leaks may create a little moisture.

Cockroaches "smell" water, and food, on certain segments of their antennae, a process that seems to become increasingly acute as the period lengthens in which they have not eaten or drunk. Entomologists have spent months of time and lots of money to show that, under normal conditions, a roach will often not react to nearby food or water. Take a roach that has been deprived of water for a few days, however, and put some water near it. The bug invariably makes a beeline for the water, running, literally a-flutter with excitement. It's really not such an earth-shaking conclusion. Humans behave in much the same way. After a full meal, we can walk by the most alluring odors without noticing them. The smell of garlic frying in olive oil or freshly baked bread stays in the background, claiming no attention, but if we are hungry, we may find that not only does the smell attract our attention, we may actually stop stock-still, turn from our intended route, and go inside to buy whatever it is we are smelling. If a cockroach is not hungry or thirsty, the message from its antenna that says "food" or "water" is just one of many sensory messages being constantly received, and not particularly noteworthy.

Also like humans, most of a cockroach's body water is in its blood, in the hemolymph. An American roach's body water content has been generally said to constitute about 70 percent of its total weight. I read this in a 1965 article from the JOURNAL OF INSECT PHYSIOLOGY, called, "Blood, Volume and Water Content of the Male American Cockroach, Periplaneta americana L.,— Methods and the Influence of Age and Starvation," in which the authors starved some cockroaches to death while daily measuring their blood/water volume, and heated others in water (when the water temperature reached 160 degrees Fahrenheit they all died) before determining the water content of their blood (Wharton, 1965). In the authors' discussion of the methods they used, there was no answer to the first question that occurred to me about their methodology: how in the world would someone go about measuring the weight of water contained in a cockroach?

"Simple. You weigh the roach, dry it in the oven, and weigh it again," said Louis Roth, when I went to see him at his Harvard University laboratory. Roth is the dean of cockroach researchers. In 1954, he and a colleague, Edwin R. Willis, wrote a ground-breaking monograph, THE REPRODUCTION OF COCKROACHES, published by the Smithsonian Institution, and in 1957 the Smithsonian also published their work, THE MEDICAL AND VETERINARY IMPORTANCE OF COCKROACHES, an exhaustive documentation of the diseases carried by roaches and some of the ways in which they might pass them. There were a number of studies in the field, but no one had ever looked at them all together to see if cockroaches really were a threat to human health. Roth's work made it clear they were.

Roth, eighty, spent his professional career as a civilian scientist working for the U.S. Army, which, like many institutions that gather numerous people together to feed and house, is plagued by roaches. He spent thirty years at the Army's Quartermaster Research and Engineering Center in Natick, Massachusetts, just outside of Boston. There is no one else whose work with cockroaches even comes close to the stature of that done by Roth and Willis. Long before the age of computers they combed through the historical record and brought together most of the relevant information about cockroaches from around the world in an exemplary feat of scholarship, and their volumes remain the standard references in the field. After Lou Roth retired from his Army lab in 1977, Harvard gave him a small office above its Museum of Comparative Zoology. When I met him in the summer of 1998, he was coming in to work every morning, getting up at 5 A.M. and driving to his lab in from his home in nearby Sherborn, Massachusetts.

"For the last few years, my work has been confined to taxonomy," said Roth, a short man with a trimmed beard and a fringe of gray and white hair. "They used to estimate that there were three or four thousand known species of cockroach. Now there are at least five thousand known and at least that many not known. It's easy to pick up new species. Not in the United States, but I get collections from different people all over the world. I can tell right away whether it's new, and there aren't too many people who can do that, so people send me odd species to be identified."

He had a modest grant from Australia that paid for his taxonomic work on samples of roaches found in Australia. His task was to identify the species to which the specimens belong. He was certainly the right person for the job. In his professional lifetime he had identified more than 350 new species of cockroach. The roaches came to him from Australia with a tag describing in detail where they were found. "Twenty miles NW of Queensland Station in leaves by Sand Creek," the handwriting on it might say, the tag tied by a piece of string around the body of a small, dead roach. Roth goes through them, one by one, examining them under his microscope. It is a solid, bulky, hefty microscope, its body made of black cast iron, the same one he has been using since the late 1940s.

"When you get a specimen, let's say a *Blatella*, if it's something found in the woods or a garden somewhere, the only way you can really determine what it is would be to use what we call a key, which is descriptive characters—you have a choice of two characters, you eliminate one, then the remaining one is paired with another and you choose between those two and you keep eliminating characters until you reach a species that has been described and that tells you what it is."

Roth opened his own mail. There was no telling what might be in it. "The roaches I get are from the wild. I'm not really interested in household pests. I've done enough with them," he

Figure 6. Madagascar hissing cockroaches *Gromphadorhina portentosa.* © Betty Faber.

said, an edge of boredom in his voice, which quickly turned to enthusiasm. "Have you ever seen some of the really big roaches?"

"No. I'd like to," I said, in the knee-jerk, bright-eyed reporter's voice that has gotten me into trouble so many times before, and was about to do so again.

"Wow," I continued in the same vein when he showed me a cage full of Madagascar hissing cockroaches *(Gromphadorhina portentosa)*, each nearly the length of my middle finger, for God's sake.

"These are the descendants of my original colony. Here, let me give you some to take with you," he said, rooting around in a cabinet and emerging with a small Tupperware container that looked like it might hold three small tomatoes. He took the top off the cage and reached in and began pulling out roaches and dropping them in the Tupperware. As he touched each one, it hissed like an angry cat.

"Are they hissing?" he asked. "I can't hear them anymore."

He wrapped his fingers around them and kept his thumb on top as he transferred seven roaches to that one small container. He lingered over the last one, turning it over in his hand and running his thumb along its underside. "This is a female and she's pregnant. I can tell."

Great. A blessed event in my future. I spluttered my feeble excuses: thanks very much but I really couldn't; I was traveling and wouldn't really be able to care for them; I would eventually be going back to Barcelona, Spain, where I lived, it would be a long trip, and I thought the authorities might be inclined to annul my hard-earned residency papers if customs agents found a plastic container full of live, giant roaches in my luggage; I appreciated the idea of the gift, it's the thought that counts; and please don't think I wasn't grateful, but they were better off staying at Harvard.

He waved away my protests. "No, no. Just leave them in the Tupperware until you get home. They'll be fine. I've got an old banana skin in the garbage around here somewhere that you can put in there. That's all they'll need. Maybe change banana skins after a couple of weeks. They'll be fine."

CHAPTER 2

the mob

TIMES WERE NOT EASY IN 1969 for a southern Jewish boy living in New York City with no skills and no college degree, so I was grateful when my friend Roy, originally from Mobile, put in a good word and landed me a job washing dishes at a club called the Scene. We had to have the glasses washed and covered with towels on top of an exposed surface, a table in the middle of the kitchen or on the bar top, when a guy came in every couple of weeks to spray for cockroaches. He should have sprayed some of the customers, instead.

The manager of the place was an intense, perpetually nervous man named Teddy. He was only thirty, but he was drawn and had deep circles under his eyes. He had a truly high-stress job. The club was in Midtown and was said to be owned by Steve Paul, a prominent name in the record business, who never set foot inside the door after dark, but occasionally came by when we were setting up in the late afternoon for a huddle with Teddy. The Scene had once been a watering hole for the likes of Jimi Hendrix and Janis Joplin when they were in town. The music industry draws organized crime like road kill draws flies, and the club had also become a favorite for a gang of a dozen or so "made" guys from Brooklyn, led by a man named Junior.

I met Junior during my first night at work. He wandered back to the kitchen. I was bent over the double sink, arms up to my elbows in soapy water. "You're new here, ain'tcha? My name's Junior, what's yours?" he asked, and I turned around to see who it was, and told him my name, nodding my head toward my wet hands as an excuse for not holding one out to shake.

Junior was young, mid-twenties, black hair combed back away from his forehead, good looking in a nervy kind of way, but there was a vacancy in his blue eyes. "Pleased to meet you. I run things here. Do what I tell you or I'll kill you." He didn't smile at all, not the ghost of a grin.

An involuntary bark of a laugh escaped me. "Are you laughing at me?" he asked, incredulous, threatening.

"No, " I said, holding my soapy offenseless hands up, palms out, in front of me. "I'm laughing because I don't know what else to do. What can you say when someone tells you that? I believe you. Hey, I don't want any trouble."

Mollified, he headed back out to take his seat at the bar. Cocaine was definitely the drug of choice among Junior's boys. There was always a lot of it around, selling for $600 an ounce in those days, and these guys were seriously out of control. No fear was the vibe they gave off. We take what we want. There is no right and wrong. Just us, lit up on toot, getting what we feel like having. Try and stop us.

Some nights they threw glasses, some nights they followed women into the restroom and terrified them, kept them trapped there while they talked at them; no rapes had been reported, but everyone knew it was only a matter of time. Other nights they got into fights and broke up the place. One night Teddy, the manager, tried to break up a fight between one of Junior's boys and a customer, and got stabbed beneath the thumb on his left hand. He sat in the kitchen with a white towel wrapped tight around it, the towel slowly turning from pink to red. When the bleeding wouldn't stop, Teddy finally took a cab to the hospital to have it sewn up.

It was after this incident that he decided to hire a *dojo*, a group of martial artists from somewhere in Queens, led by a guy named Skipper who hailed from Kingston, Jamaica. These were all big guys, square-built, dark-skinned, no-nonsense Latins, Jamaicans, and Asians. They had crew cuts, extremely serious demeanors, and looked as if they spent every spare minute working on their bodies. Three of them came in one afternoon and staked out bar stools for themselves. Then they each hammered a nail halfway into the bar by their chosen stools. Dojo members spent the nights on those stools, rotating the duty between them, with soft drinks in front of them, driving their fists into the exposed nail heads to toughen the calluses around their knuckles. It was impressive stuff. At least for me.

By then, I had been promoted out of the kitchen to second barman and was glad to have the extra security of that trio on the other side of the bar while I worked. Junior's gang, on the other hand, was singularly unimpressed by the dojo members. Within a week they had tested the new security force. The results were inconclusive. I was off that night. It was just as well. Although no weapons were drawn, the place got torn up so badly that we spent the whole next afternoon sweeping up broken glass. Tension increased around the bar, night by night, as it became clear that a more serious confrontation between the two groups was inevitable.

Despite it all, the place was open each night until 3 A.M., just another hip club with a big bouncer at the door controlling a velvet rope, deciding who would be allowed entrance and who would be rejected. The money rolled in from the regulars and wanna-be regulars, and each night we locked the doors with Teddy at 4 A.M. and went down the street for something to eat at a Greek-owned joint called the Famous. I was usually so tired by the time they brought the food to the table that I was already disconnecting, unwired enough so that my appetite had been replaced by exhaustion. I would put my order in a take-out box and head downtown to sleep. I left the food in the refrigerator and reheated it the next day before coming into work.

One of Junior's boys was a reader. A guy named Tommy, who was also the only one who wore glasses and spoke with any care for the English language. Often, early in the evenings, before everything and everyone got wound up, he would sit at the bar nursing a drink and we would talk about books. He was a big John Steinbeck fan, OF MICE AND MEN was his favorite. Tommy was a vacuum cleaner when it came to the coke, and every so often would drink himself into a murderous, cocaine-fueled fury. He had already served prison time twice for felony assaults. One more fall and he would go down for life under New York state's "three strikes" sentencing policy.

One evening around midnight, when I was pouring drinks, Tommy got into it with Skipper, the head of the dojo, who was sitting next to him at the bar. They started yelling at each other, talking trash. Someone said something on Skipper's other side, and he turned his head for a moment to look over his shoulder. Tommy's hand came up and scooped a highball glass off the bar and broke it quietly against a leg of the stool on which he was sitting. Then he stood up with the jagged glass held in one hand behind his back. He took a step toward Skipper. Without thinking, I reached across the bar and lifted the broken glass out of his hand and tossed it in the trash barrel behind me, all in one move.

He turned to see who had done it, and his look scared me cold all the way down my windpipe to my belly. "I can't believe what you just did," he said, low and menacingly.

Say something, I urged myself. "Tommy, you know it means life inside if you lose it like that. Things wouldn't be the same without you here." He looked at me a minute more and turned away. From that moment on, I knew I was alive at his sufferance. Fortunately, the situation didn't last long. The club closed a few weeks later, and my time serving the mob was done. At about the same time, the Manhattan district attorney began investigating some transactions involving Junior and his boys, and they rapidly began to be found either dead (if thought to be inclined to provide information to the D.A.) or on trial for a number of serious felonies. Most of them were eventually sentenced to substantial time upstate, according to what I read in the newspaper.

. . ● ●

After mob boss John Gotti was sent to prison for life in 1992 by the testimony of a former employee who used to kill people on Gotti's orders, a NEW YORK TIMES article quoted the capo from prison, calling the informant, "a cockroach." Wherever the mob turncoat, Salvatore ("Sammy the Bull") Gravano, was when he first read that, his scrotum must have tightened a little, because in addition to being a reflection of his character, it told him he could be eliminated as easily as squashing a bug, and that this was the fate that awaited him.

According to the OXFORD ENGLISH DICTIONARY, the first written use of the word was by Captain John Smith, Pocahontas's lover, who wrote about cockroaches in 1624, in his book, THE GENERALL HISTORIE OF THE BERMUDAS: "A certaine India Bug, called by the Spaniards a cacarootch, the which creeping into Chests they eat and defile with their ill-sented dung. . . ."

Pocahontas, however, may have misjudged her lover's character. Smith has been accused (Kevan, 1981) of plagiarizing most of his book from Nathaniel Butler, who was appointed governor of Bermuda in 1619, and immediately began taking and publishing copious notes on the island's fauna, including the following: ". . .the moscitoes and flies are somewhat ouer busie, with a certain Indian bugge called, by Spanish appelation, a caca-roche, the which, creepeinge into chestes and boxes, eate and defile with their dung (and thence their Spanish name) all they meet with . . ." (Caca, of course, being Spanish for feces.)

While cockroaches are the world's oldest surviving insect form, the word, cockroach, is a relative newcomer. The

Greeks knew them as *blats*, and the Romans called them *lucifuga*, for their habit of avoiding light. The insect was prevalent in the Old World, and the Oriental cockroach, *Blatta orientalis*, was a common pest in medieval English homes, but the word, *cockroach*, did not come into usage until explorers began traveling to other continents. It seems to have turned up in order to denominate a new species, not the old familiar black beetle, or stinking moth, or *blatta*, all names by which the resident English species was commonly known, but rather something that came back aboard the explorers' ships when they returned from their wanderings through Africa, Asia, and the New World. In the mid-1500s, the word appears in a work by Spanish playwright Lope de Rueda as an insulting phrase, *cucaracha de sotano* (basement cockroach). At approximately the same time, it appears in German and Dutch as *kakerlak*, in Creole French as *coquerache*, and in Parisian French as *canquerlin* (in addition to the French *cafard*, which was already in use for the resident French *Blatta orientalis* and is also used in the slang phrase, *J'ai le cafard*, which means I'm feeling a little sad, or I've got the blues).

The word *cockroach* also has other dimensions than the one simply describing an insect; nowadays, it can carry the psychic force of a serious curse, an epithet of deep hatred, which is how John Gotti certainly meant to use it. A review of the references to cockroaches in the popular press, the similes and metaphors in which the word is used, quickly reveals that around the world, in the most disparate of cultures, to call someone a cockroach is to express an utter contempt for the individual in question, to deny any value whatsoever to a life, and to imply the threat of a raised shoe.

In the 1994 slaughter of Tutsis by Hutus in Rwanda, Hutu radio broadcasts and propaganda constantly referred to the Tutsis as cockroaches. Before sentencing former Rwandan prime minister Jean Kambanda to life in prison for genocide, a United Nations tribunal heard a string of witnesses recall how Kambanda handed out weapons to Hutu civilians and urged them to hunt down and kill "Tutsi cockroaches." The epithet was not a new one. A Tutsi woman in Rwanda told NEW YORKER writer Philip Gourevitch about seeking refuge as a teenage girl during the 1973 massacres of Tutsis by Hutus. She walked weeks to reach the home of a female Tutsi relative who was married to a Hutu man. He turned the young woman away, saying, "I don't give shelter to cockroaches."

In 1983, the chief of staff for the Israeli Army, Rafael Eitan, was asked about how Palestinians would react to a planned increase in the number of West Bank Jewish settlements. "Like drugged cockroaches in a bottle," he told the NEW YORK TIMES, causing a bitter outcry among Palestinians.

"Drug organizations are like cockroaches," New York City Police Chief Howard Safir told a press conference in 1996. "We need to spray them occasionally."

In short, if you want to say something nasty about someone, call him a cockroach: that lowest of the low, vilest of the vile, most easily eliminated without a pang of remorse, the cheapest of all lives, an animal only a Jainist sworn to respect life in all its myriad manifestations—who wears a mask to save microbes from being swallowed and lightly sweeps the earth ahead of him as he walks to brush tiny creatures aside out of harm's way—would ever think twice about killing. Roaches are synonymous with dirt. They are among the most repugnant of life-forms. A Manhattan exterminator told me: "In most of the buildings I go to, there are people who won't let you into their apartments. If it's a guy, it could mean that he has a kilo of heroin on the coffee table, or an AK-47 in the corner, or something illegal that he doesn't want you to see, but if it's an older lady, the reason she won't let you come in and spray, or lay out bait for the roaches, is that she doesn't want to admit she *has* roaches. Just having the occasional roach means to her that there's something to be ashamed of, that she doesn't keep as clean an apartment as she should."

While it's true that roaches are not paragons of cleanliness as we think of it—they will walk through and even eat, if they're hungry enough, their own excrement, and have no hesitation about feeding on the most repulsive substances including their dead colleagues and live offspring—they are also, in their fashion, as obsessed with cleanliness as the most immaculate of housekeepers. For instance, they spend an inordinate amount of time each day grooming themselves. They keep their antennae clean by hooking them with a front leg and pulling them through their mouths in an almost feline manner.

. . • ●

Now Sam watched Sharon, who had heard the calling of her name twice, followed by a long silence. Sam's sniffwhips detected an unmistakable scent of yearning. The bristles on the lower tip of sniffwhip segments are especially sensitive to scents of yearning, wanting, inexpressible wistfulness, in either one's fellow roosterroaches or in Man, and it is considered good luck to pick up, on one's sniffwhips, such a pining smell . . .

How does a fastidious genteel roosterroach know when his nightly (or thrice-nightly) ablutions are finished? Of the 178 segments on each sniffwhip, the last two, at the very tip, have as their sole function an appraisal of one's own cleanliness, tidiness, and aroma. Sam no less than any other roosterroach would rather have lost both his tailprongs and been totally deaf than to lose the tips of his sniffwhips. Whenever an individual loses these, through accident, battle, or failure to keep them clean, that individual is almost certain to be dirty, stinking, and flowzy . . . until he regenerates the tips.

For all man's repugnance toward him, the roosterroach is the most immaculate of insects, permitting no speck of dirt or disease to remain upon his body. And Gregor Samsa Ingledew was the most immaculate of rooster-roaches.

from *THE COCKROACHES OF STAY MORE*
by Donald Harington

•   •   •   .

The first book of Donald Harington's I read was *SOME OTHER PLACE. THE RIGHT PLACE.,* which came out in 1974, was a wonderful book with a huge reach. Harington's capacity to create a wide variety of characters is amply showcased in the array of novels he has gone on to write, in all different voices, textures, and times, most of them set in the woods and mountains of northeastern Arkansas. *THE COCKROACHES OF STAY MORE* chronicles the lives of two communities in a small Arkansas town—the *Blattarian* and the *Homo sapien.* Written from an insect's point of view, the story traces the effects of love on both humans and cockroaches.

A cockroach deals with the world through its antennae. The antennae may be longer than the rest of the roach, depending on the species. A tremendous part of a cockroach's world has to do with the sensory reception of chemical signals, a sense that for humans is most closely allied with that of smell. Sentience may in fact be made up of only five senses in every species, but the way they manifest has a wonderful variation. The roach—despite its hard shell and generally bad repute as a feelingless, insensitive piece of protoplasm—is designed with some remarkably sensitive equipment, much more so than our own. The roach's senses are finely honed on a level we can hardly imagine (unless we are gifted with an imagination like Donald Harington's). Entomologists tend to play the wonder of insect antennae down—simple receptors for chemical signals, they will say—and over the years cockroaches have had their antennae stimulated, deadened, amputated, grafted, injected, radiated, and dissected. Despite all that slicing and dicing, nowhere in the literature that I found does it specify just how many segments there actually are on antennae. Most would say Harington's figure of 178 is probably a little high, with many putting the average number at about 135, and still others at 150. In fact, it appears likely that antennae have no fixed number of segments but vary, to some extent, from animal to animal (Schafer, 1973). For all the attention paid to its composition and development in the laboratory, the cockroach antennae is a foreign country that remains to be accurately mapped.

What is known is that the antennae are covered with fine hairs—cilia—and connected by a nerve to a part of the cockroach brain that both sends and receives signals, so that incoming neural messages from the antennae can be translated on-the-spot into action, and the brain can also send out messages to control the movement of the antennae. Those incoming neural messages, how can we conceive of them? What does the nearby presence of water translate into on a cockroach antennae? Or food? Or receptive females? In the case of olfactory function it is believed that there are receptors on the antenna's sillae designed to receive particular odor molecules (Norris, 1976). These receptor cells are made to fit only a limited number of molecules. Some, for instance, respond to alcohols and not to sugars, and vice versa. Foods like meats and fruits combine a variety of chemicals and they stimulate a number of different cells on the antennae. Males have a substantially larger number of olfactory receptors than females, and it is believed that these extra cells are for receiving pheromone molecules, the smells that tell a male when a female is receptive to his sexual advances.

In addition to functioning as an olfactory receptor, the antennae also have cells with thermoreceptors for measuring cold and heat, and others that register humidity. And, they do part of the job that in humans is done by eyes. The cockroach

Figure 7. A cockroach's head. © Betty Faber.

has a multifaceted eye, made up of about 2,000 tiny octagonal lenses, plus, in most species, a "simple" eye for detecting light and darkness. The insect does not see as we do. From its compound eye it receives a mosaic vision, the acuity of which is not known, but it is not believed to be particularly sharp, especially when there is light. A roach seems to depend more on its antennae than its eyes to examine its immediate environment, and the antennae move almost constantly, like the white cane of a blind person walking, exploring what lies ahead and around, although in addition to bumping up against things, the antennae transmit a host of other sensory data.

A pair of neuroscientists working at the University of Illinois found that if you touch a *Periplaneta americana* on its antennae, it will almost invariably turn away from the touch (Ye, 1996). They did not, to be sure, simply touch an American cockroach on an antenna and see if it turned the other way. Instead, they designed a motion tracking system that measured the roach's neuronal activity using a specially designed electrode implanted at the cervical connective. The cockroaches were anesthetized by being chilled at 4 degrees Celsius, then the electrode was inserted. The roach was given an hour to recover, and when it was moving again, experiments were run using the motion tracking system. The research was aimed at measuring three distinct aspects of escape behavior: how long after being touched on the antenna it took a roach to initiate escape behavior, the angle of the initial turn away from the stimulus, and the trajectory of its entire escape run.

"There are very few instances in which it has been possible to describe the signaling of specific nerve cells in relation to natural behavioral responses on a trial-by-trial basis. In vertebrates there are a few cases in which the firing of classes of cortical neurons has been correlated with behavioral decisions.... In 90 to 95 percent of behavioral trials, intact animals turned away from the sides on which an antenna had been tapped.... How is this initial direction of turn established? A major outcome of this work was to show that, when one antenna is touched, there is a lateralization of... descending impulse activity in the large DMIs [descending mechanosensory interneurons]...." wrote Shuping Ye and Christopher Comer in their discussion of the experiments in the JOURNAL OF NEUROSCIENCE. Of course, if a cockroach could read, the most disturbing phrase in the above material would be *intact animals,* and, in fact, the neuroscientists found that cutting the neural pathway resulted, not surprisingly, in the roach frequently turning toward rather than away from the touched antenna.

For all their sensitivity to stimuli, the antennae apparently can only function at close range. Put a cockroach more than a few inches away from food or water and it doesn't seem to register its presence. For an animal as vulnerable as a pest cockroach, and as apt to excite a hostile response from at least some components of it surroundings, such as humans, it is imperative to be able to sense danger at greater distances than just a few inches. By the time a raised shoe is within a few inches of a cockroach's body, it is usually too late. A roach's early warning system is located not on its antennae, but on its backside, by its anus, where a pair of feelers called cerci, erect and tapered— wide at the base, narrow at the top and considerably shorter than antennae—are covered with fine hairs, hundreds of them, 0.5 millimeters long and 0.005 millimeters wide, each with its own single nerve fiber. These hairs are remarkably sensitive, and serve as motion detectors. They are the reason that when a person turns the light on in a roach-infested kitchen at night, by the time the room is illuminated, the cockroaches are already running full tilt for their hidey-holes. It calls to mind the great Negro league baseball player, James "Cool Papa" Bell, famous for his speed around the base paths, whose roommate said he was so fast that when he turned off the light in their room at night, he was in bed before it got dark. Likewise with a roach, except the insect doesn't just seem to be that fast, it *is* that fast. Just by entering a room, a person sets air in motion, even before reaching for the light switch, and the cerci register those air currents before the lights come on. When a cockroach feels a breeze stirring the hairs on its cerci, it does not wait around to see what is going to happen next. It leaves off whatever it is doing and goes immediately into escape mode in something remarkably close to instantaneous fashion. The main signal-carrying nerve of the cockroach is the ventral nerve cord, which runs like a spinal column the length of the body from down by the anus where it forks and connects to each cercus, all the way up to the brain. When the hairs on the cerci are set in motion by moving air, a signal is sent through a nerve in the cercus directly to the leg muscles along the ventral nerve, without having to first be routed to the brain. Experiments have shown that a decapitated cockroach will still start running as soon as air is blown on its cerci. It takes a cockroach about one-twentieth of a second to react when something riffles those cerci hairs, so in less time than the blink of a human eye it is in motion. By the time a light comes on, and human sight can register it, much less react by reaching for and hoisting something with which to squash it, a roach is already locomoting toward safety. And, safety is likely to be close at hand, because a roach's flattened shape allows it to squeeze under, through, or behind just about anything. In addition, apart from its sophisticated motion detecting equipment, a roach is just plain fast. They run about fifty body lengths a minute, which for a human being would be the equivalent of two hundred miles per hour.

In addition to carrying the signal, which is an electrical nerve impulse that causes the animal to react, the large ventral nerve also carries chemical messages that *do* go to the brain, and there they initiate the production process for a variety of hormones. Cockroaches actually have two brains—one inside their skulls, and a second, more primitive brain that is back near their abdomen, a meeting point for principal nerve fibers that has the capacity to store information. The more sophisticated brain is in the head, however, and consists of three parts—one that is neurally connected to the eyes, a second to the antennae, and a third to the nervous system in the rest of the roach's body.

The nervous system is not the only means by which chemical signals are transmitted—they also circulate in the hemolymph and are responsible for the regulation of organ function, and for what Cornwell calls "activity cycles." The idea that hormones and chemicals function as regulators and initiators of certain kinds of activity was proposed and proven by Berta Scharrer, one of roachdom's most distinguished researchers, who died in 1995 at the age of eighty-nine. Her work with the species *Leucophaea maderae* disproved forever the old notion that a cockroach's nervous system was nothing more than a series of wires and connections for transmitting sensations and impulses. She and her husband Ernst Scharrer both believed the nervous system controlled the secretion of hormones, and devoted their working lives to proving their theory of neurosecretion— he working with vertebrates, and she with insects, a natural division since she had received her graduate training in Germany under zoologist Karl von Frisch, who won a Nobel prize for his work with bees.

Scharrer and her husband decided to leave Germany as Hitler consolidated power. Even though they were not Jews, they had Jewish friends and could tell that they did not want to be around for what was coming. They arrived in Chicago in 1937, each with a suitcase and eight dollars between them. Her husband received a one-year fellowship at the University of Chicago, where Berta was able to use a little corner of a laboratory, but she did not even have money to buy lab animals with which to carry on her work. When a janitor showed her the American cockroach infestation in the basement of the lab building, she saw not pest but providence. Her husband's next post was at Rockefeller University in New York. American cockroaches were plentiful there as well, but when Scharrer discovered a group of *Leucophaea maderae* stowed away in a shipment of lab monkeys from South America, she began working with them and kept on doing so throughout her career. The *Leucophaea* has become a popular lab animal, because in spite of its strong cockroach odor, it is larger than the American roach and easy to handle. Even though Scharrer did not receive her first academic appointment until 1952, by the time her husband died in 1965 she was receiving worldwide attention for her innovative work. Eventually, Scharrer settled in at Albert Einstein University in the Bronx, and maintained her lab there until she retired in 1993. Her work with the *Leucophaea* and the hormones it produces gave birth to an entirely new discipline: neuroendocrinology, and in 1983 she was nominated for a Nobel prize.

While a cockroach's ventral nerve carries signals back and forth to its brain, the cells of that big, central nerve, itself, also contain substantial quantities of hormones, among which is bursicon, a peptide hormone, that is necessary to harden a cockroach's shell after each of its molts. Bursicon is produced only at this time and is only present in a cockroach's hemolymph for a matter of hours after it molts. Cockroaches are members of the phylum, arthropod, which includes other insects, and animals like crayfish and crabs, that also grow by molting. The only time that a cockroach can grow is when it molts. It discards its old skin, the tough exoskeleton called chitin, by gulping air until its shell splits, crawling out from it and gulping air again until it forms a new, larger shell, allowing the developing roach inside, called a nymph, room to grow for a while before its next molt. At each molt, it doubles its body weight. The number of molts from birth to adulthood depends on the species, but almost all have at least half a dozen. In the case of the German cockroach, there are usually six molts and they happen within about sixty days.

The molt is, obviously, one of the most important processes in the life of any roach. This is how the animal proceeds toward adulthood and reproductive capability. Exactly how the process works is still unknown, but what is clear is that it is initiated several days before the actual shucking off of the old exoskeleton. The American cockroach, for instance, molts about

Figure 8. A newly molted cockroach. © Betty Faber.

every twenty-eight days, and by day twenty-four the process is already beginning, with the new shell forming beneath the old one, along with a change in chemical composition of the hemolymph. Eventually, when the nymph has split its old exoskeleton and stepped out of it with a new, larger shell into which it will grow, it will eat the molted shell, recycling the chitin and its proteins.

This is also when a nymph can replace any limbs that it has been unfortunate enough to lose since its last molt. Legs and antennae will regenerate in the new molt, as will cerci. There are some fourteen breaking points on the legs, cerci, and antennae of German cockroaches, so that if they are grabbed by a predator they can pull away and leave the enemy holding nothing more than an appendage, which will be replaced at the next molt. Once a cockroach becomes an adult, it loses the ability to regenerate limbs.

It does not take a cockroach long to begin molting—its first one will occur within five days of hatching out. The stages of this process are known as instars, so that entomologists refer to a nymph that has gone through two molts as a third instar cockroach. A newly molted cockroach will spend its first couple of hours as a pale, almost transparent dot, as its body produces enough of the hormone bursicon to begin to darken it and harden its exoskeleton. Many an unsuspecting home owner will come on these newly molted roaches and put them in a jar, thinking they have captured an albino cockroach, only to find, a few hours later, that they have turned into standard-model roaches.

A new-to-New-York-City resident, a young woman named Marlene Matarese, who had come from the Midwest to study environmental design, suffered the same sort of acclimation to sharing her living space with German cockroaches as many another midwesterner come to live in Manhattan. It must have been a slow news day at the NEW YORK TIMES when her letter about it came in, because they sent a reporter to talk to her and listen to her story about how bad things had been at the apartment she shared with two other students when she first came to town. It had been completely infested with cockroaches. "We used so much spray that we had to leave the apartment," she told the newspaper in 1977. "But, it didn't do anything to the roaches. We had so many in the kitchen that the sink was black at night. They were so confident they didn't even move when we turned on the light. It was frightening. We had such a gene pool of roaches that we were getting mutations. We had albino roaches."

Well, not quite. There are albino roaches, but they are no more frequent in the general population than is the case with albino humans. What Marlene thought were albinos were, no doubt, simply newly molted cockroaches that would have stayed pale for the two hours it took their bodies to produce enough bursicon to darken their developing exoskeletons.

Willi Honegger, a biologist from Munich, Germany, who teaches at Vanderbilt University in Nashville, figures that if he can just disrupt the process by which bursicon molecules bind to their receptors and harden the new exoskeleton, it will result in a permanently soft-shelled cockroach, which, in turn, will prove fatal to the insect in short order. In service to this hypothesis, he has extracted bursicon from the main ventral nerves of over 6,000 American cockroaches.

I went to see him one summer morning at his Vanderbilt laboratory. It was mid-June in Nashville, and graduation was over. Summer school had not yet begun and the campus was deserted. The science building was vacant and silent; its library was empty, waiting, a solitary librarian perched on a stool behind the counter reading a paperback; big lecture halls were unoccupied, their chalkboards clean; antique brass scientific instruments were displayed for no one in tall old wooden cases with high glass fronts in the abandoned air-conditioned halls.

"Bursicon is found in the nerve cells, and we take these nervous systems right out and homogenize them," Willi Honegger, an older, balding man with a gray/white fringe of hair around his poll and a thick German accent, told me when we met in his laboratory. "We have a source for free, live cockroaches. It's a company. They are interested in our research, but I'm sure they would prefer I not tell you their name. When the cockroaches come in, we put them in the freezer and when we need them we thaw them, and take the nervous systems out. From that we extract the bursicon."

The we to whom he refers are biology students, who get a little financial help with their stiff Vanderbilt tuitions, as well as practical experience by working with him. That summer morning he had Missy Williams working in the lab, a sophomore from North Carolina who was bent over a "teaching" microscope with two sets of eyepieces. She turned away from hers to greet me, then turned back and invited me to join her by watching through the other. She was lovely, tall with short blonde hair, long legs, and flawless skin, wearing the tiniest of turquoise minidresses with much more of her exposed than covered. I obediently put my eye to the microscope and watched as she pinned a roach on its back in a petri dish half filled with a liquid she told me was identical to cockroach hemolymph in its chemical composition. She cut the roach up the middle of its abdomen with a tiny pair of scissors, delicately laying back the flaps of segmented belly to reveal a white thread nestled in the body cavity of the roach and running from its hindquarters to its head. This was the main ventral nerve, and she began delicately snipping it away from the body with the tiny scissors, using a wee tweezers to lift it away as she cut, until out it came. She transferred it to a second petri dish of hemolymphlike fluid where it floated with others.

Missy narrated as she worked. "I'm cutting the muscle around the nerve cord and the first three ganglia, now. At first I didn't realize a cockroach had so much fat. Like, it's disgusting at first. I'm studying, like, molecular biology, genetics."

Everything under the microscope is reversed by its mirrors. So what I see as her left hand is actually her right. Both her hands worked quickly, expertly. "When I first started doing this and saw that everything was reversed I thought, like, gosh, it'll take me years to learn, but I caught on, like, pretty quickly," she told me, all the while looking through the microscope and tweezing away at the nerve, until she expertly freed and lifted it out. It was amazing to my layman's eyes how the little dish and the roach pinned on its back became an entire, full-field, watery world under the lens.

It's going to take a lot more ventral nerves for his research to bear fruit, according to Honegger. "We need to find something that can compete with bursicon for the two hours after the molt in which it's active. It's very difficult. We might have something in another four or five years."

. . ● ●

One thing that surprised me about traveling around with a small Tupperware container holding Lou Roth's seven huge hissing cockroaches was how profoundly they disturbed people. Not everyone disliked them. Some folks do, after all, purchase and keep them as pets, but the great majority of people absolutely cannot seem to abide being around them. My brother, for instance, a guy who sails his thirty-foot Pearson Wanderer into the teeth of storms with hardly a second thought, despised them at first glance. I was staying at his coastal Connecticut home when I drove up to Harvard and interviewed Lou Roth. My rational and brave brother went more than a little crazy when I returned from Cambridge bearing the bugs, singularly unimpressed when I explained to him that they were a gift from an *eminence gris* of roach research, fresh from his own colony, and that as big as they were, the Madagascar hisser had not been born that was strong enough to push the top off the Tupperware from the inside. Nor did he seem to believe me when I told him that they sold for $6 a pair in exotic pet shops around the country. When I mentioned that I wanted to leave them at his house for a couple of weeks while I traveled around doing my own roach research, my brother initially refused outright, finally consenting but insisting that I tape up the container with duct tape and leave it on a high shelf in the back of his garage. He seemed to shudder just thinking about the roaches, and did not like it when I brought them up in conversation. He was not alone. His friends who visited, male and female, did not even want to come out to the garage and look at them, although my young nephews, of course, were more than willing to hold the Tupperware up to the light with me. On the eve of my departure, I peeled away the tape, opened it, and pushed one of the Madagascars with my finger so they could hear it hiss. My brother made me go to the alley behind his house so I wouldn't even be on his property when the top came off. When I put the tape back on and restashed the Tupperware on his garage shelf, I knew I couldn't count on any fraternal help in changing the banana skin, or even checking to see if they were still alive while I was gone. And, I have to admit that some substantial part of me was hoping that the next time I opened that container, in a couple of weeks' time, it would hold seven corpses. Then, I wouldn't have to think about how to explain to a customs agent, in a language not my own, that I had no idea how a sealed plastic bowl full of giant cockroaches had gotten into my suitcase.

la zona

IT WAS DRIZZLING BUT NOT COLD in Ciudad Juarez, when I found myself working on a story there in early December 1997. I was privileged to pass a day with one of the city's true believers, Graciela de la Rosa. She spent her mornings working at a women's center for low wages, then went back to her shadowy, high-ceilinged home for a long vegetarian lunch with bread she had baked herself in this house where she had grown up, which had been built in the 1920s. Graciela had inherited the house from her parents and had opened it up to a communal arrangement with three other people, a trio of like-minded individuals who she told me felt as she did: the only way to live in this sad and sorry world is to dedicate one's days to living modestly and helping others. In Ciudad Juarez there was no lack of folks around in need of help.

Juarez is across the Rio Bravo (or Rio Grande, depending on which country is doing the naming) from El Paso at the western end of Texas. It is a boom town of a sort, experiencing the global economy's equivalent of gold-rush growth. Juarez has two huge industrial parks, and they are chock-a-block with *maquiladoras*, factories set up by multinational corporations to assemble components brought from the States into products that will be exported, free of taxes, back to first-world countries to be sold. This takes advantage of the large pool of cheap labor that exists in Mexico and creates huge profit margins for the multi-nationals by drastically reducing wages from what they would have to pay if they made the products where they sell them. The Mexican government, trying to cope with an unemployment rate that is estimated to hover around 40 percent, was glad to have them. The *maquilas* principally hire women because they are more dexterous. The workers are paid about thirty-five dollars for a forty-eight-hour work week on their feet doing things like wiring a circuit board, or sitting down sewing a sleeve on a shirt, making the same movements all day with demanding quotas to meet if they want to keep their jobs. In many of the *maquiladoras*, women must bring in bloody proof of their menstrual periods each month to show that they are not with child. Pregnancy is cause for dismissal.

Most of the foremen in the plants are just that: men. Women do the repetitive motion tasks because they are quicker and more nimble with their hands, more productive. There is full employment in Juarez and that is an amazing thing in a nation where there are many citizens who simply have no work, where there is nothing for many people to do and no social safety net, where tens of thousands of people daily go undernourished for lack of the most meager work at which to trade their time for enough to feed their families. Day and night the radio stations in Juarez run ad after ad announcing which plants are hiring, and tall signs are posted outside many *maquiladoras* announcing which shifts have vacancies. Many of the factories run around the clock. They want to hire people under thirty—and this is in a country where it is common to see young men sitting under trees alongside the road, looking into a hopeless middle distance. These young men are the real meaning of an "underdeveloped country." They are the euphemism, *underemployment*, made flesh and bone. They sit watching traffic go by, waiting for nothing by roadsides, their lives running out just like yours or mine, no possibility of paying work, no flocks to tend nor ground in which to plant, nor any labor to do for pay. These men do not know how or what or whether they and their families will eat that night. Men and women all over Mexico, willing and eager to work, are sitting in the doorways of daub-and-wattle homes with thatched roofs, because they have absolutely nowhere else to go. But not in Ciudad Juarez. There the radio stations are broadcasting offers of immediate employment at better than minimum wage on the night shift at the General Motors or Phillips or Sony *maquiladoras*. This, in turn, has drawn tremendous numbers of people from small villages in the south, and Juarez has nothing approaching the infrastructure needed to cope with the influx.

There are 1.2 million people in Ciudad Juarez, of whom 200,000 are employed by a *maquila*. In fact, the unemployment rate there is 1.2 percent, and that of El Paso, Texas, across the river, is ten times higher at 12 percent. Of course, in El Paso, even the desperate would never work under *maquila* conditions. First shift, for instance, is 6 A.M. to 3:30 P.M. with fifteen minutes for breakfast at 8 A.M., twenty minutes for a midday lunch, and one permitted bathroom break, all for less than ten dollars a day.

With so many women working and so many families dependent on the seven or eight dollars that a wife or daughter brings home each day, there are entire villages to the south in which the primary source of revenue is money sent from Juarez. It is a much more pernicious immigration, in many ways, than illegally crossing the border to the States and working. Mexicans in the States can more easily retain their cultural identity, they still feel they belong back there, in the village, in the traditional life. They are only abroad in a foreign country to earn money, and will sooner or later be back in the old ways, the strong ways, as soon as they have saved enough money to allow them to return. Not necessarily so for those people who have remade their lives in a border town so they can work in *maquilas*. For them, the old ways are gone, they are still in Mexico, but it is a Mexico where the women work and live on their own or with each other, spending some part of their own earnings as they please. Gone are the youngsters married as teenagers and committed for life to making babies and tending to family. Gone are the birthing, living, and dying amidst one's own relatives, the strong and widespread family-based culture that is such an important part of Mexican life. These women have left that sustaining familiarity and relocated to Juarez where their closest friends may be *comadres* from the same village, or they may have no one with whom they can bond and be entirely on their own. Border culture can be dangerous for these women, physically as well as psychologically. Since 1995, over 140 young women working in Juarez's *maquiladoras* have turned up dead, their violated, and sometimes mutilated, corpses found in the scrub brush outside of town. A young, solitary woman makes easy prey and, as of 1999, when Juarez officials formally requested help from the FBI in solving the crimes, it was not known whether this was the work of the worst serial killer in history, or a number of men.

Graciela works closely with women in the *maquiladoras*, and she says some of them turn to prostitution to earn a wage with which they can feed a family, or to augment their income from the factory. A survey she directed for the women's center questioned 207 prostitutes in Juarez, and 40.5 percent of them had worked at *maquilas* before taking to the street. They had changed professions, from one of the world's newest to one of its oldest, and over 60 percent had made the switch for what they described as "economic" reasons. Some 12 percent had done so because of family problems, which we know are often exacerbated by poverty, and 7 percent because they were bored. There is a whole other group of women who works in the *maquilas* during the week and as prostitutes during the weekend.

The hookers work out of the tiny cantinas that flourish in *la zona*, the part of downtown Ciudad Juarez where there are more small bars than stores. The broad boulevard is packed with people and lined with places to order a shot of mescal across a rough-hewn board. Cassette tapes of *rancheros* and *cumbias* play on an old boom box cassette/radio. It is also the part of town where many of the young women who turned up murdered were last seen alive.

Down the middle of the boulevard are little seafood kiosks with three or four stools, where the principal dish they offer is ceviche. I would sooner swallow live coals than subject my digestive system to marinated raw fish from one of these places. What stores there are all seem to be shoe stores, small with brightly lit windows full of cheap footwear. Heavily made-up prostitutes stand in the doorways of little cantinas, watching the business of the streets transpire, smoking, talking, and laughing among themselves. As the setting sun shadows the street, the rouged women in the doorways look as hard as nails and the men patronizing them have a dark, furtive, desperate appearance.

The light rain has made the pavement in *la zona* slick and muddy with crushed vegetable matter and whatever else. Graciela makes her rounds chatting with the women, encouraging them to come this evening to the meeting with Father Antonio where we ourselves later go. Father Antonio is a young, soft-spoken priest with gentle good looks, who ministers to the members of a parish hard against the walls of one of Juarez's largest industrial parks. His parishioners live there, outside the walls surrounding the *maquilas,* in a *colonia* of cardboard and tin houses. A worker in one of the *maquiladoras* inside the walls may support a half dozen family members on one wage. They will all live in two rooms and count themselves fortunate. The only running water is from a handful of pipes stuck in the ground. An executive wheeling his Mercedes out the factory gates at night can hardly help but see the neighborhood where his workers live.

Father Antonio comes from the *colonia* to *la zona* once a week to bring tai chi and Bible study to the women of the street. He treats them as a loving brother would his sisters. He is easygoing and relaxed with them and seems to genuinely enjoy their company as they do his. Each grants the other dominion in their chosen field, the priest demurring to the women in the ways of the world, and they to him in the ways of the Lord. The meeting took place in a back street, close by *la zona*, in the home of a hooker named María.

María lived in two rooms with cinder block walls and cement floors. One of the rooms was a living/bedroom, and the other a kitchen with an old stove and older refrigerator. There were a half dozen parakeets in small cages hanging in corners. Also in residence were four young kittens and their mother, a gray cat the color of cinders, ears raw from where she had rubbed them for mites, nipples pink and chapped in the mouths of the kittens that looked like four tiny piles of ashes. Their mother lay on the linoleum like a rag dropped on the kitchen floor, and there were two brindle curs tied up in the alley behind the house, their ribs showing through their hides. Cockroaches ran along the wall behind the refrigerator, every so often one would swerve out onto the lit part of the wall where the kittens batted at it and the roach, realizing its error, recoiled from the light and dashed for the safety of the shadows behind the ancient rattling Kelvinator. The parakeets trilled, the mother cat meowed in a complaining voice, and the dogs barked out in the alleyway. A third room in the back of the house was rented to another hooker for one hundred pesos (twelve dollars) a week. It had an iron single bed, a sink in the corner, and a cheap bureau with a page out of a movie magazine pasted on it, which showed movie star Antonio Banderas holding hands with his wife, actress Melanie Griffith.

It was here where a rickety card table had been set up in the middle of this woman's one rented room, around which sat five hookers, Graciela, and Father Antonio. The evening I joined them, the discussion focused on the expectation of a new life, not only as revealed in the resurrection of Jesus, but in the chariot coming for Elijah. Two of the women had brought well-worn Bibles, which were shared among the group. After the evening's discussion, María told Father Antonio she thought it would be better to stop doing the tai chi and spend the time saved on Bible study. Besides, she was not sure whether the exercises might be aggravating her asthma. Graciela convinced her to stick with the tai chi a little longer. Graciela felt the tai chi was every bit as important as the Bible study, and led the women, in their short skirts and bare feet, through the exercises.

I asked to use María's tiny bathroom. She looked embarrassed as she led me to it. "It's there," she pointed to a door in one corner, "but don't flush the toilet because the water's not running right now. It quit working this morning in the whole neighborhood."

Many *maquiladoras* operate under conditions that neither labor nor environmental law would allow in the United States. At the other end of the long Texas border from Juarez, Brownsville sits just a walk across a bridge from Matamoros, which is Mexico's easternmost city. In 1995, *maquilas* on the Mexican side agreed to settle a lawsuit against them for $20 million. The suit, brought by some twenty-seven Brownsville families, accused forty-five Matamoros *maquiladoras* of having discharged toxic wastes into the air, where they were blown by prevailing winds across the Rio Grande, provoking a rash of birth defects in Brownsville, particularly spina bifida, which prevents babies' spines from fusing, and anacephalia, which causes babies to be born without brains. While the settlement specifies that the multinationals do not admit guilt, the Harlingen, Texas, lawyer who represented the families, Randy Whittington, assured me there would never have been a settlement if the companies were innocent.

The *maquiladoras* have brought employment to the border, but more than that they have not not offered. Despite the fact that they are making tremendous profits, the *maquilas* rarely leave any of that money behind, except in the form of small paychecks. Cities like Matamoros and Juarez have had their populations increase exponentially over the past twenty years, but since the multinationals have invested practically nothing in local urban infrastructures, problems with things like housing, waste treatment, schools, roads, and public transportation have multiplied.

The plants have provided lots of jobs over the years, but they have paid little attention to improving the communities in which they find themselves. Profits, like the products assembled at the *maquilas*, get shipped back north, according to environmental activist Domingo Gonzalez. He is the son of migrant workers, who grew up working in the fields of California. As a young man, he organized with César Chávez in the farmworker's union, and has spent thirty years as an activist. For the past decade, he has lived in either Brownsville or Matamoros, "working on environmental issues," which makes a person about as welcome in some border circles as cockroaches at a formal dinner party. Domingo Gonzalez receives a small stipend from the Quakers to raise his unpopular voice, and his wife works in a *maquiladora*.

"We've seen the *maquilas* arrive and turn the border towns into huge belts of misery," he told me. He was walking on crutches, and had been for more than a year. His right leg was crushed in a car accident when a pick-up truck ran him off a road in Matamoros and sped away. His leg gave him considerable pain even when he was sitting down. He had no doubt that someone had intended to silence him by killing him.

"Normally, when you're someplace where a factory opens with jobs for three, or four, or five thousand people you say, 'That's great.' Streets get built, highways, theaters, and big hotels. That hasn't happened on the border. Instead of new highways, airports, and grand hotels, what we have are enormous belts of poverty, and tremendous problems in controlling what happens to our environment."

. • • ●

Down in *la zona* in Ciudad Juarez a slim woman in blue jeans and a denim shirt passed me by going in the opposite direction. Her hair was fine and inky black. I glimpsed eyes as dark and deep and soft as any I had ever seen. I crossed the street through her back draft, opened my nose and inhaled something that fell between gardenias and sweat, an odor all hers, still there hanging in the air for a moment before it was volatilized and subsumed in the larger smell of *la zona*, which was compounded of garbage and cigarette smoke, ceviche and cheap perfume, and God knows what else. Children breathe on their arms and sniff the vaporous humid warmth that carries the odor of their own skins. The lingering smell of that woman's passage was as pleasant as the smell of my own arm at seven years old under the Tennessee summer sun.

.   .   •   •

Pheromones, chemical signals of sexual readiness, operate between a male and a female cockroach to initiate courtship and copulation. A sexually receptive female assumes a posture with her abdomen lowered and her wings raised, and gives off a pheromone that attracts males. The production of the pheromone, and female sexual receptivity, itself, is dependent on a hormone produced by the corpora allata, a large gland behind the brain (Scharrer, 1946). In some species, the pheromone is a volatile chemical that is dispersed through the air and will bring a male cockroach from across a room, while in others, the male only senses the pheromone when he touches the female with his antennae. Louis Roth, and his colleague Edwin Willis at the U.S. Army's research laboratory, spent years studying sex among American cockroaches, and their forty-nine page volume, THE REPRODUCTION OF COCKROACHES, published by the Smithsonian Institution in 1954, brought together the work already done by others with their own. It is the classic reference volume on the subject. Nevertheless, at first the Army was more embarrassed than pleased by their work. The two scientists originally named their monograph, THE SEX LIFE OF COCKROACHES, and, in fact, submitted it to their Army superiors under that title.

Too risqué, decided the top brass. "Don't forget, this was the early 1950s," Roth laughed, telling me the story. "You didn't talk about sex, even among roaches."

In this case, the contents were more important than the packaging, and the work soon became a basic text in the field. The two researchers were the first to demonstrate that *Periplaneta americana* females give off a sex attractant. They put virgin, unmated adult females in containers lined with filter paper. The paper absorbed enough of the pheromone, so that when a male was placed near the paper he displayed courtship behavior. Roth and Willis determined that within a few hours after her final molt into adulthood an American cockroach is producing pheromones, and this production increases for about two weeks, after which it tapers off.

They were also the first to describe the German cockroach's mating behavior, and observed over 6,000 instances of copulation between members of this species. Until then, the sex life of domestic roaches had not been recorded, and this material became increasingly important as more scientific extermination strategies were developed. There are many ways to kill roaches other than squashing them, but only a limited number of routes by which to attract a roach to come within range of a death-dealing substance. In order to deliver a deadly dose of any insecticide, regardless of how it will kill the bug, advantage must be taken of a roach's habits and needs, which are relatively few and revolve—like our own, for that matter—around food, shelter, and sex.

It is, after all, not only roaches that are vulnerable to the Siren's song of pheromones. Human beings smell each other consciously and subconsciously, and experimenters have raised the possibility that a broad range of our own reproductive behaviors are based on the subconscious reception of smells. Armpit sweat, saliva, and urine contain several known pheromones, and experiments have shown that both men and women respond behaviorally to these odors without consciously registering them, and that they also play a role in the onset of puberty and the regulation of the female menstrual cycle. Children can identify the sex of a person from their smell, and mothers can pick out their own newborn in a hospital nursery, simply by smell (Kohl, 1995).

One pair of researchers described the reaction of the male American cockroach to the sex pheromone: "... the interest of the male ... manifested first by its sudden alertness, then movement of the antennae, then active search for the source of the odor, then more or less vigorous fluttering of the wings, usually accompanied by protrusion of the abdomen" (Wharton, 1957). The normal reproductive process for *Periplaneta americana* is relatively simple. When the male senses the pheromone, he turns around and flutters his wings, raising them as an invitation for the female to mount him and begin the long copulatory process.

In German cockroaches, the kitchen species, things take a little longer to get the male going. The German female does not release an airborne attractant, but has a pheromone in the grease on the cuticle of her exoskeleton that lets the male know she is receptive. He must make physical contact to be stimulated. A male who meets up with another roach will touch it with his antennae, and if he finds a virgin female, after a brief bout of antennae rubbing, the male will turn away from her and raise his wings, an invitation for her to mount. The rapid stroking of each other's antennae has been proven to be an indispensable first step in the *Blattella germanica* mating ritual, and experiments have shown that when male cockroaches are forcibly separated from a receptive female immediately following an antennae stroking session, they retain some sort of pheromone, because if that male is subsequently placed in the company of other males, they, too, become sexually stimulated. There has to be a pretty strong aphrodisiac in play here, because normally cockroaches seem to have no trouble distinguishing between sexes. In fact, one roach can tell if another roach is male or female simply by a mutual stroking of antennae. Roth and Willis cut the antennae off male and female roaches, and when they stroked the antennae of other roaches with the severed organs, the stroked roaches responded differently depending on whether it was being done with an antenna that had been severed from a male or female, and their responses were always accurate.

In both species, a willing female will respond to the male's raised wings by climbing aboard him. The female German cockroach, and many other species mount the male slowly, nuzzling his back, feeding on a substance that is produced specifically for courtship by a pair of glands under the male's wings. Without this love-meal, a female cockroach will not proceed with mating. She will avidly lick the male clean as she climbs up on his back, in a process that it is not hard to imagine provides pleasure for both of them, until she has fully mounted him. Once the female is atop the male, he maneuvers his genitals into contact with hers, first using one of his two phallomeres, a gristly sort of hook-shaped appendage, usually concealed beneath the male's shell, which stretches, extending as he becomes aroused, probing inside the female and, in the case of a successful mating, hooking on to a sclerite he finds there, a sort of half circle of muscle just inside her genital chamber. (In the first English textbook devoted entirely to the cockroach, the male's hook was called, aptly enough, a "titillator" [Miall and Denny, 1886].) Once they are successfully hooked together, he twists around and out from under her in such a way that they are back end to back end, facing away from each other. Copulation frequently lasts an hour.

.   .   •   •

E.: "Wait. Have a little patience. Don't try to go so fast. Put my hand in some place that gives you pleasure. Come on, let's sit on the sofa, okay? You underneath because I love to put my naked ass on your legs." A. followed to the sofa, asking

Figure 9. Antennae fencing. © Betty Faber.

Figure 10. Wing-raising posture of the sexually receptive male cockroach. © Betty Faber.

Figure 11. The male cockroach inviting the female to mount.
© Betty Faber.

Figure 12. End-to-end copulation. © Betty Faber.

all his cockroaches to put themselves in his sex, that they leave his eyes and his heart and his hands, that they were needed down there...

A.: "My body's asleep. The only thing that wakes me is to think of the disappearance of C. and B. I'll ask D. to hurt them a little before she kills them. So they feel pain. That they understand what torture is..." In effect, the cockroaches were gathering in a group where they had been called.

E.: "Ah!" E. felt those cockroaches as they moved on her ass and felt an indescribable pleasure. She wanted them inside. She showed them the route with a hand. They had a hard shell, shiny and slippery. E. wanted to put them in her mouth and squeeze them between her tongue and her palate until they let go of that white juice that comes out when they are squashed. She didn't do it because, in that moment, she didn't want to be left without a cockroach.

A.: "Yes, I'll ask D. to make them take their clothes off and..."–in that precise moment of thinking of his wife and her lover without clothes, all A.'s cockroaches, which had begun to enthusiastically go up and down inside E., suddenly went back to the eyes and heart of A. without any warning.

E.: "What's happening to you? What happened?" –disappointed, she knew that now neither the tongue nor the palate would serve for anything.

A.: "Is it really so important, now?" he answered, impatiently. He was there without clothes on as if he were in front of the doctors who had examined him in the military.

from the prize-winning 1997 Catalan novel,
NI TU, NI JO, NI NINGÚ
(NOT YOU, NOT I, NOT ANYONE)
by Flavia Company
translated with the author

● ● ● ·

As in almost every other aspect of their lives, the evolutionary strategy employed by cockroaches to reproduce is considerably more efficient than that of humans. The internal genital organs of male cockroaches include testes, seminal vesicles, and accessory genital glands. From these latter comes a secretion that is used to mold a spermatophore, a carrying case for the spermatozoa, to be passed over to the female. The most active secretion occurs within the first few hours of the emergence of the adult male following its sixth instar molt. He then retains the chalky-white secretions in his gland ducts. If a male goes months without copulating, the ducts swell and swell with this substance. It is not hard to imagine that in this condition he would be primed to find a receptive female, and would spend a good deal of his time looking for one.

As soon as he successfully hooks up with a female and initiates the copulatory process, a male expels these chalky-white secretions into his ejaculatory pouch, where it will surround sperm that has been delivered to the pouch from the testes. The secretions will harden to form a sperm packet, and inside this spermatophore will be two sperm sacs. Once formed, the spermatophore descends the ejaculatory duct. During copulation, a male passes over his sperm packet to the female where it enters her genital pouch. She stores the sperm and uses them as needed, and they can last her a lifetime.

Scientists have observed sperm from an American cockroach used nearly a year after it was passed to a female (Griffiths, 1942). This, in itself, is obviously a tremendous design improvement over the human blueprint. For us, fertile men and women must always have access to each other, and each copulation normally has the potential to produce only one more human. This reproductive advantage guarantees roaches a thriving population. Adding insult to injury, there are even certain species, including *Periplaneta americana,* that are capable of reproducing, at least for a generation or two, by parthenogenesis, which means that the female's unfertilized eggs will develop and hatch—always producing new females—without any sperm.

Once the male has passed his sperm packet, his job is done. Mission accomplished. The spermatophore enters the female and comes to rest with its two openings, one for either sperm sac, right at the paired openings of two large glands in the female called spermatheca, which is where she will store the sperm. Once the sperm packet is positioned in front of the spermatheca, she activates the sperm with a chemical message, and they begin their journey into her sperm bank. In the German cockroach, the transfer of sperm usually takes about twelve hours, after which the female expels the empty sperm packet and leaves it behind, going on her way with a lifetime's supply of sperm safely stored in her body.

Much of a female cockroach's adult life is spent pregnant and taking care of the business of reproduction. Males seem to lead a somewhat more carefree existence, according to Betty Faber, one of the nation's experts on cockroaches. She is the staff entomologist at the Liberty Science Center just outside New York City in Jersey City, New Jersey, a stone's throw from Ellis Island and the Statue of Liberty. Every day during the summer the center is filled with kids from day camps who crowd noisily around the exhibits, including a big cage teeming with Madagascar hissing cockroaches.

Faber is a short, solid, single mother with trimmed blonde hair. An entomologist's entomologist, she was wearing silver earrings in the shape of spiders the day I visited her, and a large silver cicada brooch. In her small office upstairs at the science center she has a gallon jar of live American cockroaches on a shelf beside her desk. "I grew up in Biloxi, Mississippi, and I was scared to death of roaches, although I was always sort of fascinated by insects. We lived out in the country. Roaches were the bane of my existence when I was growing up. In the summers I wouldn't go out to the carport to get my bicycle because there'd be big old roaches in there. I went to college at William and Mary and was scared of roaches there too, but when I came up here to do graduate work in circadian rhythms, how animals tell time, there was an opportunity to work with them. One of the hardest things to learn was to put my hand into a jar like this and pick out a roach. . . . "

She reached over to the shelf beside her desk, unscrewed the top from the gallon jar and dipped her hand in among the scurrying American roaches, pulling one out to show me, then casually replacing it among its fellows. "I had to force myself to learn, because if you do it with forceps, it just doesn't work. Do it too hard and you kill the roach. Grab a roach by the leg with forceps and it just drops the leg, and gets along fine on four or five legs. I found that if I used forceps I was not going to have time for a social life or any other existence, so that's how I learned to pick up roaches. It took me seven months. Believe me, that was really scary."

I did not have any trouble believing her. I was trying to sharpen my own skills as a cockroach handler, and it was not easy. I had to keep consciously admonishing myself to pick them up, not flinch. If I relaxed my vigilance for an instant, something deeper took over, urging me to drop the bug as if it were hot. Most people, on a primal level, do not want to make direct contact with a cockroach.

Before coming to the science center, Betty Faber was the staff entomologist for the American Museum of Natural History in New York. After eight years of commuting from Princeton, New Jersey, to Manhattan proved too much, she took the post at the science center because it was closer to home. The years she spent at the museum were productive ones. There is a greenhouse situated on the museum's roof that contains "an excellent specimen colony of naturally occurring American cockroaches," and Faber spent many nights during her tenure at the museum observing that colony. Cockroaches are most active at night, and the best tool for watching them is a flashlight with a red gel over the light. Roaches cannot see red and for them such a flashlight is the same as no light at all. Before Betty Faber's seven years of fieldwork atop the museum, there were almost no reports on cockroach behavior outside of a laboratory setting. Most university researchers are reluctant to submit themselves to the rigors of spending their nights observing *Blattarians* in the wild, even if it is nothing wilder than a greenhouse over the Museum of Natural History.

Faber labeled individuals in the museum's colony with numbers written in Magic Marker on bits of adhesive tape stuck to the thorax and wings of her subjects, and she would make the rounds of the greenhouse every couple of hours with her flashlight to see who was where doing what with whom. One thing she discovered was that males will consistently stay out later than females. "Females go to bed—by which I mean disappear back to the harborage—at night earlier than the males. In the greenhouse you'd rarely see a roach during the day, but in the early part of the evening you'd see a hundred at a time. Every couple of hours I'd walk around the greenhouse and write down where I saw the roaches and what I saw them doing. I found out that by midnight most of the females would have disappeared, but the males would stay out until two or three. In the morning they're all gone.

"I'm not 100 percent sure why, but if a male only has to mate with a female one time to make her pregnant for life, then in terms of biology and evolution, once he's done that a couple of times he's not needed anymore for carrying on the species. If anybody's just going to hang around, or look for food, or do whatever they do, it's not so bad that he does it. They're there when they're needed. In the early part of the summer when the females are producing pheromones and the males start their courtship dance, you can hear it. You may not be able to see them, but you can hear them. The pitter-patter of tiny feet," she laughed.

To many people, the only thing uglier than a German cockroach is a pregnant German cockroach, her ootheca sticking out her back end. As repulsive as this sight may be to the layperson, it has long fascinated scientists, and by the end of the nineteenth century the ootheca had already been well studied. This little, ridged, hard-shelled carrying case for the fertilized eggs is made of glandular secretions that form what Miall and Denny, in 1886, called "a mould of the vulva," and then harden almost to the toughness of plastic. A female German roach may produce as many as forty oothecae during a lifetime of one year, and each one could hold two rows of twenty eggs, as many as forty eggs per ootheca (Roth and Willis, 1954). As the ootheca is forming in the German cockroach's vulva, she deposits two rows of eggs in it. As the rows of eggs lengthen, so does the ootheca, until it begins to stick out, extrude as the entomologists would have it, from her body. Anyone who occasionally sights a German cockroach is likely, sooner or later, to catch a glimpse of one trailing an ootheca, or find an egg case dropped somewhere in the vicinity of where roaches are usually seen.

A female cockroach produces mature eggs within a few hours of emerging from her sixth instar, and continues to do so for most of the rest of her life. The eggs move out of the female's ovary, past the spermatheca where they are fertilized by the stored sperm, and on to the ootheca. When they leave the ovary, the end containing what will be the nymph's head is pointing up toward the mother's head, and the eggs are aligned in the ootheca with the head end up toward the top of the egg case, where it will eventually split open. If the head is not pointed up, the nymph will develop normally in the ootheca but will not be able to hatch (Roth and Willis, 1954). The ootheca, itself, is a marvelous construction, designed to offer the right amount of humidity and

Figure 13. A female cockroach with an egg case sticking out of her back end.
Photo © Louay Henry.

correct temperature for the fertilized eggs to develop. In the ootheca of the German cockroach, for instance, there is a little space left above the compartment for each egg that has a microscopic opening to the outer air.

Among the thousands of species of cockroaches, ovipositional behavior—what a female roach does with her eggs—is subject to considerable variation. Some carry extruded oothecae around like the German and drop them just before they hatch, and others fill the oothecae with fertile eggs and leave them behind, going on their ways with never a backward glance. The American cockroach is in the latter group. She will produce about eight oothecae per month to be filled with eggs and will find a dark place in which to leave them, often covering them with leaves or dirt or debris, sometimes chewing up such material into a paste to spread around an ootheca in order to keep it in place. That is the greatest degree to which she will exhibit maternal concern.

Other species form and fill an ootheca that is then retracted back into the body where the eggs develop and either hatch in the mother's brood sac, resulting in a live birth of the nymphs, or are expelled shortly before the nymphs emerge. Roth and Willis timed one female member of the species *Pycnoscelus surinamensis* and found she completed the entire process of making an ootheca and fully retracting it, loaded with eggs, in two hours and fifteen minutes. In contrast, a member of *Leucophaea maderae* took five-and-one-half hours to do the same thing. Fossilized cockroaches from prehistoric times tell us that some ancient roaches were equipped with a long ovipositor that apparently laid single eggs, one by one, without an ootheca, while others were equipped with the same sort of short ovipositor as today's roaches, indicating they deposited their eggs in a carrying case. There is an exception to the current rule of the ootheca. It has been reported that *Diloptera punctata* actually carries live, developing young in her brood sac without an ootheca, and that she produces a milklike substance, high in protein, carbohydrates, and fats to nourish the young as they grow inside her. Of course, *Diloptera punctata* is an odd one, anyhow, a wood- and leaf-eating cockroach found in great numbers in Hawaii. When attacked, it sprays a strong-smelling liquid at its aggressor from its tracheal gland (Eisner, 1938).

The amount of time required for eggs to hatch depends on temperature. At seventy-five degrees it will take a German cockroach nearly thirty days, but at eighty-six degrees the time is cut to seventeen days (Rust, 1995). When the nymphs are ready to emerge, which can be anywhere from fifteen to ninety days in the ootheca, depending on the species, they begin swallowing air, increasing their size, and forcing the ootheca open along its keel. Sometimes, it will not rupture, and all the nymphs will die inside, or it may open and snap shut again before all the nymphs have emerged, killing those left behind. Within a few days, the nymphs that successfully leave the ootheca will undergo their first molt and be on their way to their own reproductive capacity, a process that will lead German cockroaches through six molts within sixty days, and into adulthood.

Not all pheromones are directly related to sexual activity. There is another that plays an important role in bringing the sexes together, and that is the so-called aggregation pheromone. It is not a sex pheromone, in that it does not stimulate a roach sexually, but it is what leads roaches to a common harborage, and, of course, the more cockroaches in a given place the more likely it is that they will find suitable mating partners.

It was A. Ledoux in France who reported in a paper published in 1945 with the title, *ÉTUDE EXPÉRIMENTALE DU GRÉ-GARISME ET DE L'INTERSATTRACTION SOCIALE CHEZ LES BLATTIDES (AN EXPERIMENTAL STUDY OF GROUPING AND SOCIAL INTERAC-TION IN THE BLATTIDES' HOUSEHOLD)*, that roaches will band together if given a chance. He postulated that there must be a chemical attractant motivating the behavior. However, it was not until the late 1960s that Japanese researchers first isolated the aggregation pheromone, and used it on filter paper to stimulate aggregative behavior in other roaches, in much the same way that Roth and Willis did with the sex pheromone. The chemical compound that stimulates cockroaches to gather together has not been deciphered beyond determining that ammonia and a baker's dozen of other amines are present (Sakuma, 1990). The aggregation pheromone is clearly distinct from those at work during courtship, and it is found in the insect's fecal matter. In 1970, Kyoto biologist Shoziro Ishii ground up numerous German cockroach rectums and used the resulting paste to impregnate filter paper. The paper drew roaches like death draws flies, and Ishii concluded that the aggregation pheromone is secreted by the rectum pads as one of the final ingredients in the feces. He also determined that if a cockroach's antennae were cut off, it would not react to filter paper impregnated with aggregation pheromone.

In an infested apartment or kitchen, in the places under shelves and around cracks where cockroaches can be observed to gather, are thin streaks of brown that might have been drawn with a pinhead dripped in brown ink. These are cockroach droppings, and newly arrived *Blattarians* need only encounter them to follow a trail toward a harborage where they will find a colony of fellow roaches. The aggregation pheromone in the fecal matter, which calls the roaches home with its chemical message, is powerful and long lasting. When fecal matter leaves the roach's body it is wrapped in a thin membrane, which appears to slowly release aggregation pheromones over a long period of time. Cockroach *caca* can still bring roaches together even after it has been exposed to the air at room temperatures for nearly a year (Metzger, 1995).

Cockroaches, while not social insects in the entomological sense that bees or ants are with clearly assigned tasks that benefit the whole community, do clearly take pleasure in the company of other roaches, and the aggregation pheromones draw them together, eliciting their effects regardless of the sex or age. In fact, for our pest *Blattarians*, optimum development depends on growing up in a community. Cockroaches reared singly develop more slowly and take longer between molts than those in a group. German entomologist Roland Metzger has suggested that information regarding surroundings is passed along in the aggregation pheromone.

.   .   •   ●

Life as life. To remove myself from it, I like to observe cockroaches. Each is boring individually but together they have great potential.

Yesterday for example. I'm sitting in the kitchen having a smoke and they're running around until, at a certain moment, they form into Leonardo da Vinci's masterpiece *Mona Lisa*. By chance? No, only the inevitable law of development, the creative dynamism of the group, evolution. It's enough that this society was running around and there are results.

The problem was that they immediately ran in all directions. Leonardo lasted only a second. I thought I'd take the bug spray, wait and, when they form again into some achievement, I'd spray and fix them in place. I clutched the spray and waited.

They started running around again. Something like Manet's *Dejeuner sur l'Herbe* flashed by for an instant. I let it pass. The cockroaches had evidently advanced in their development and were seemingly in the impressionist period. I could have frozen them there but do I have the right to halt their development? Impressionism is a major achievement but who knows what they would achieve?

Cubists—I let them pass.

Surrealists—I also let them go.

I held my finger on the spray but I didn't press the nozzle. I know something newer must come after the new, that is, after good comes better. Not to worry that Leonardo and the later ones ran away. On the contrary, that's progress.

We're in the contemporary period now. Masterpieces as well. Warhol, for example. But he's not the last word. He's already a classic. Keep running, kids, and run into something that never was. I'm waiting for the most up-to-date modernity, for the best.

What's that! I don't see anything, only cockroaches running around. Are they tired? Some sort of decadence? The downfall of art? I stared, but there was nothing but cockroaches.

How dumb I am! How can I see something when I haven't developed yet? They must be in the twenty-fifth century (because it was already past midnight and they were running fast) and I'm still in the twentieth. My perception has lagged behind, that's all.

I put the spray away and went to bed. I'll go back to the kitchen in five hundred years.

"Evolution" by Slawomir Mrozek
published in *TYGODNIK POWSZECHNY*
(No. 43, 1988)
translated from the Polish by Michael Sulick

● ● ● ·

There is, of course, always the danger that the aggregation pheromone will create a situation in which there is too much of a good thing, and living quarters become too crowded. Just as development is delayed in young cockroaches if they are isolated, overcrowding also extends the time between molts. To counter the ill effects of crowding, there is yet another kind of pheromone, called a dispersal pheromone, and it serves as a chemical signal that it is time to look for a new, slightly roomier harborage. This chemical is found in the insects' saliva (Suto, 1981) and has just the opposite effect of the aggregation attractant, in that it repulses cockroaches and causes them to look elsewhere for harborage. It *is most prevalent where overcrowding exists.* Filter paper impregnated with saliva from roaches living under such conditions repels other roaches that come in contact with it. It is the chemical equivalent of a no trespassing sign. While researchers hope to one day utilize such dispersal pheromones in their ongoing battle of the bugs, dispersion from one site often leads to infestation close by, and this is not usually a desirable outcome in roach control.

The various pheromones have been found on cockroach bodies, in their fecal matter and saliva, and volatilized in the air.

These chemical messengers are received by specialized receptors on the antennae and prompt specific social and sexual behaviors. It seems reasonable to assume that there are other messages in this chemical language of which we are not yet aware but that contribute to an insect's daily perceptions of its world.

． ． ● ●

"How're the cockroaches doing?" I asked my brother on the way back to his house from the train station.

"How would I know?" he asked sharply, in the same offended voice with which he always seemed to respond when the subject of the Madagascar hissers came up. "They're dead or alive."

I wasn't sure which I preferred, but it turned out that two weeks in cramped quarters on a back shelf of his garage with one banana skin had no evident ill effects on the roaches, who clambered around as soon as I peeled off the duct tape and opened the lid. I could see that there was a sizable accumulation of tiny, dark brown feces, and chose not to think about it. I had other more weighty issues to consider, such as how I would get them home to Barcelona.

Hand luggage was out of the question. What if one wiggled an antenna going through the X-ray machine? What would the security guard at Kennedy Airport say after picking *that* up on the monitor? One thing was certain: there was no need to worry that the radiation from the X rays might harm the bugs. Cockroaches survived the atomic bomb's test blast at Bikini Island in the Pacific, Lou Roth had told me. There is such a thing as a lethal dose of radiation for a cockroach, but it is a lot higher than our own. Regardless, I didn't want to chance questions from security, so hand luggage was out.

That meant they would have to go in my baggage and withstand an eight-hour, trans-Atlantic flight in the freezing hold of a plane. Not exactly an easy way to make the journey for heat-loving cockroaches that originally hailed from a subtropical island. Nevertheless, *Blattarians* have historically proven that they are perfectly capable of withstanding the rigors of overseas travel, and they have been doing so for centuries. Their preferred method of intercontinental travel has traditionally been on ships. Like rats, cockroaches came to the New World as stowaways, and the most persistent pest species are thought to have arrived aboard slave ships (Rehn, 1945). It is, perhaps, fitting that such an evil commerce as slavery provided American and German cockroaches an opportunity to invade our hemisphere and establish themselves. Slavery and syphilis were two societal ills brought from the Old World that would poison the blood of the New, even unto the present day, and cost untold numbers of lives. With them came domestic cockroaches to take up residence in our homes.

Cockroaches have, apparently, always been at home on ships. It is widely accepted that both the German and the Oriental cockroach traveled from North Africa in Phoenician or Greek vessels to Byzantium, Asia Minor, and the region of the Black Sea. From Russia they went to Western Europe and on to America. Wherever men went in ships, it is safe to assume that cockroaches went with them. Cockroaches were an accepted and normal part of seagoing journeys, so much so that they often did not warrant serious discussion by the chroniclers of shipboard life. They do turn up, however, mentioned in passing again and again, in the records of voyages, many of which were collected by the indefatigable Roth and Willis in their BIOTIC ASSOCIATIONS OF COCKROACHES, published in 1960. The earliest record they found was Thomas Mouffet's THE THEATRE OF INSECTS: OR, LESSER LIVING CREATURES, first published in Latin in 1634, and in an English translation in 1658. Mouffet noted that when Sir Francis Drake captured the Spanish vessel SAN FELIPE in 1587, a galleon loaded with spices, it was found to be infested with cockroaches.

If cockroaches were required to pay for passage on ships, the German would take a cabin, while the American would settle for steerage, because preferences for exactly which part of the vessel to infest varies according to species. The German is likely to be found in the galley, or the crews' quarters, or the officers' cabins, while the American cockroach usually chooses to travel in the hold, or near the steam pipes in the bowels of the ship.

Both species seem to get on and off ships pretty much at will. The British naturalist, Henry Nottidge Mosely, sailed around the world between 1872 and 1876 on the H.M.S. CHALLENGER. He noted that the ship began to have cockroach problems after calling at the Cape Verde islands. One particular bug became the bane of Mosely's existence: "One huge winged Cockroach baffled me in my attempts to get rid of him for a long time. I could not discover his retreat. At night he came out and rested on my book-shelf, at the foot of my bed, swaying his antennae to and fro, and watching me closely. If I reached out my hand from bed, to get a stick, or raised my book to throw it at him, he dropped at once on the deck, and was forthwith out of harm's way.

"He bothered me much, because when my light was out, he had a familiar habit of coming to sip the moisture from my face and lips, which was decidedly unpleasant and awoke me often from a doze. I believe it was with this object that he watched me before I went to sleep."

Strategies for treating the problem of roaches on board have been as varied, and generally unsuccessful, as those employed by homeowners on land. The infamous Captain Bligh, whose crew on the H.M.S. BOUNTY later mutinied against his tyranny, wrote in the ship's log in 1792: "This morning, I ordered all the chests to be taken on shore, and the inside of the ship to be washed with boiling water, to kill the cockroaches. We were constantly obliged to be at great pains to keep the ship clear of vermin...."

Although there is no record of a bounty being paid on the BOUNTY, a system of rewards has been frequently tried as an extermination measure. In the Danish navy, during the early 1600s, sailors received a bottle of brandy for every thousand roaches caught. Three hundred years later, in the early 1900s, Japanese seamen who captured as few as three hundred were given a special one-day shore leave.

In addition, Frank Cowan, writing in the mid-1800s, records that monkeys and lemurs were often kept on board ships to dine on roaches and keep their numbers down. He quotes an article from an 1829 issue of the MAGAZINE OF NATURAL HISTORY, in which one P. Neill describes bringing a marmoset monkey back to Scotland from Brazil: "By chance we observed it devouring a large Cockroach, which it had caught running along the deck of the vessel; and from this time to nearly the end of the voyage, a space of four or five weeks, it fed almost exclusively on these insects, and contributed most effectively to rid the vessel of them. It frequently ate a score of the largest kind, which are from two to two-and-a-half inches long, and a very great number of the smaller ones, three or four times in the course of a day. It was quite amusing to see it at its meal. When it had got hold of one of the largest Cockroaches, he held it in its fore-paws, and then invariably nipped the head off first; he then pulled out the viscera and cast them aside, and devoured the rest of the body, rejecting the dry elytra and wings, and also the legs of the insect, which are covered with short, stiff bristles. The small Cockroaches he ate without such fastidious nicety."

The shipboard situation has not greatly improved with the passage of time. At the turn of the twentieth century one author reported that the crews of ships crossing the Pacific slept with gloves on in order to avoid having their fingernails nibbled by roaches. In the 1930s, records from New York indicate that ships docking there occasionally carried tremendous infestations. On board one vessel there were 20,000 German cockroaches killed in a single stateroom, and 50,000 in the forecastle. Even today, cockroach infestations occur on board ships, although cruise ships take stringent precautions against them.

Many freighters are infested, and smaller boats also have their problems. There are facilities at places like the Panama Canal that will de-infest private boats, although it is a time-consuming and expensive extermination, in which the boat is enshrouded and cyanide gas bombs are set off inside it.

In a letter about the infestation of a private sailing vessel, which may reveal a possible reason for my brother's hostility toward my Madagascar hissers, he wrote me: "In Buena Ventura, Colombia, sleeping on the SPUD, a thirty-six-foot old-fashioned ketch made of inch-and-a-quarter planking taken out of a single huge Port Orford cedar in Oregon, the roaches had free run of the ship after the tropical sun went down. We more or less got used to them, but sometimes at night they would drop off the underside of the deck where they ran upside down and would land on you, only to scurry off in a hurry across your face. Nasty way to wake up. Always gone before you were awake enough to get them. Captain used to say the only way to get rid of them was go through the canal, where for a fee they would seal up all openings on the boat and pump her full of cyanide gas for twenty-four hours. We'd spend idle hours, of which there were many, as we tried to sail south along the Pacific coast of Colombia against the Humbolt current and the prevailing winds, such as they were, discussing what you could or could not safely leave on the boat, dishes, clothes, etc., during the extermination. Course, it was moot as we were heading the opposite direction, albeit slowly."

Air travel has always appealed more to many travelers than ocean crossings, and cockroaches have been boarding planes for a long time too. By 1935, public health authorities in France were warning of accidental cockroach importation by airplane. In the year between July 1, 1956, and June 30, 1957, specimens from eighteen different species of cockroach were collected on airplanes coming into the Miami airport, according to Roth and Willis, but they do not report whether the animals were collected in the baggage compartment or from the passenger section of the plane. Standard procedure is to spray the passenger and kitchen sections on a regular basis, but it is not foolproof. For instance, a 1993 Delta flight from Houston to San Francisco was delayed more than three hours after a "handful" of German cockroaches was sighted near the food service area. The company found another aircraft to make the flight.

I was not up for carrying my six-legged charges on board so they could have in-cabin comforts. That meant consigning them to the hold. A study done in 1958 seemed to indicate German cockroaches could not survive in the low-humidity, unpressurized and unheated cargo holds of jet aircraft. Roaches were loaded on jet fighters and bombers, and it was determined that in a three-hour flight at 40,000 feet in a jet bomber, with outside air temperatures as cold as fifty-one degrees below zero Centigrade, there were no survivors.

The research, however, was far from conclusive. Live roaches had been found in the cargo holds of numerous aircraft. After reviewing the literature, I decided to pack my Tupperware inside a checked suitcase. I was convinced that they wouldn't be the first roaches to survive such a trip. Still, most of the reports dealt with German, Oriental, and American cockroaches, pest species, not Madagascar hissers, who, left to their own devices, have nothing to do with humans or their machines, preferring the jungles of their native island where they feed on rotting wood. It was a real possibility that the next time I saw my roaches, they would be dead-on-arrival. Ah, well. I put a new banana skin inside the Tupperware and gift wrapped the container, reasoning that even if the customs officers stopped me and had me open my suitcase, there was a good chance they wouldn't ask me to unwrap a gift. What I would say if they asked me to unwrap it was something about which I tried not to think.

hambre

IT'S THE SMELL THAT HITS YOU FIRST, even more than the visual squalor, a smell full of rot and the heat of decomposing organic matter. There are no windows or doors left in the abandoned, crumbling brick building in the center of the huge *Mercado Oriental* in Managua, Nicaragua, the largest market in Central America. The brick ruins are alongside the market's refuse dump, a square block of open garbage disposal. The gang of street kids who occupy the building scavenge the dump beside it for food. They commit small crimes in the two square miles of the market's precincts to provide themselves with funds to buy the cobbler's glue to which they are addicted.

The smell of the occupied ruins is compounded by layers of garbage ground into the muddy floor; hundreds of pots of discarded dump meat and vegetables boiled over trash fires; and the smell of human waste from the corners where the residents defecate. As strong as it is, a smell is just a smell, and after a little while the brain no longer registers it.

"*Hambre*," María Teresa said, when she woke up one April morning in 1996, in a dazed, still half asleep voice. Bringing the fingertips on one hand together, she brought them to her lips as if carrying a bite of rice to her mouth. She had a surprised, lost look in her eyes, as she gazed up at the cameraman filming her, and me standing beside him. "*Hambre,*" she said again. *Hunger.* As though she was not accustomed to waking up hungry.

It was 8 A.M., and maybe half of the dozen or so pieces of flattened cardboard boxes that served these juveniles as mattresses were still occupied. Bodies were flung down in postures of absolute weariness. It was difficult to imagine how exhausted one would have to be to stretch out on that floor. But there they were, sleeping in the same clothes—ragged T-shirts and torn shorts—they had worn the day before and would wear again today. Barefoot. Beside each sleeping juvenile lay an empty, glass, baby food jar with a small patch of dried glue at the bottom.

María Teresa had the prepubescent body of an eleven-year-old, with the postpubescent head of the sixteen-year-old she was. Toluene (tol-u-een), the solvent in the glue that got her high, also disrupted the growth of her body. Kids who begin using it early do not complete their physical development. Her black hair was chopped off at the neck in a rough haircut given, judging from the look of it, by someone who had a glue buzz going and was using a pair of dull scissors. Her body may have been that of a child, but it had received an adult's share of punishment. Later, she pulled down the top of her T-shirt and showed me a thick scar just above her breastbone where she said she was stabbed by a policeman. The bodies of most of these adolescents bore the ragged scars of stabbings, or the puckered, ruined skin where a bullet wound had been rudimentarily repaired and left to heal. And, as bad as the boys have it, the girls have it much worse. There is something erotically exciting to some men about a woman's head on a child's body, and these girls are sexual fodder for any man who can overpower them, or pay them the twenty cents that will be enough to buy a couple of bottles of glue as soon as he finishes.

Within the boundaries of the *Mercado Oriental* are scores of curbside cobblers who use the glue to attach soles to shoes. Nicaragua's citizens are too poor to throw out a pair of shoes because they fall apart or the sole wears through. A cobbler charges less, by far, to repair the old pair than it would cost to buy new ones. Cobblers buy the shoe glue by the gallon and transfer it into baby food jars to prevent it from evaporating and drying out. The cobblers sit at small workbenches, one next to the other, in certain areas of the market, each with a bunch of the baby food bottles beside him. Some of them also have a business on the side selling the bottles to street kids at one *cordoba* (ten cents) a bottle.

María reaches for the baby food bottle on the floor beside where she has slept and sucks on it, breathing deeply, but it is as futile as she knew it would be. The bottle holds no pleasure. It still bears a trace of the characteristic sweet smell of shoe glue, but the toluene in it has long since evaporated. People who inhale glue get off from exactly that—the evaporating toluene, a petroleum-based chemical solvent widely used in shoe glues and paint thinners. It induces a kind of high, and mutes physical sensations like hunger and cold, things that these outcasts know all too well. Toluene also extracts a terrible toll on the central nervous system. In additon to inhibiting the growth of addicted youngsters, it eventually slurs their speech, robs them of the ability to walk, and, finally, kills them by causing organ failure in their early twenties. There are no old glue addicts.

No one knows how many people under eighteen live in the streets of Managua, but they estimate at least a few thousand. Of these, well over half are reported by social workers to be regular *huele pegas* (glue sniffers). An addicted fifteen-year-old can sniff six bottles a day. If you multiply six by, say, three thousand kids, at one *cordoba* a bottle, in the second poorest country in Central America and the Caribbean (Haiti being the first), this qualifies as a serious market, and there are a number of companies that have stepped up to meet the demand.

Cobbler's glue is not particularly difficult to manufacture, if one has the requisite capital to rent a factory, purchase the chemical components and the equipment, and hire some labor.

The street cobblers buy the one-, three-, or five-gallon cans with which to fill their baby food bottles from hardware stores. Hardware stores buy from wholesalers who sell the glue out of fifty-five gallon barrels that they have bought directly from the glue's manufacturers. In addition to the glue manufacturing companies formed by Central American entrepreneurs in Honduras, Nicaragua, and Guatemala, there are also multinationals profiting from the market's young glue addicts. One such is the North American paint and glue manufacturer, H.B. Fuller, Inc., headquartered in Minneapolis, Minnesota, which has long drawn heat from child advocates in Latin America for continuing to sell toluene- and hexachloraphene-based shoe glues in places where it is reaching young addicts. They have refused to modify the glue, as other companies have done, by adding a tiny part of oil of mustard to make it unpalatable for sniffing.

The Illinois-based Testor's Corporation, for instance, began adding oil of mustard to its model airplane glues during the late 1960s when young North Americans began buying them to sniff and get high. H.B. Fuller contends that to make such modifications would lessen the product's effectiveness for its legitimate users, although other glue manufacturers, including Testor's, reject this. Fuller's products continue to be abused throughout Latin America. This was the story I had come to Managua to tell with a documentary team from Catalan public television, and it was for this reason we were watching as María woke up hungry.

María's boyfriend is twenty-two, and he calls himself *el toro* (the bull). He is the leader of the gang. He is good looking in a wild, unkempt sort of way, with flashing brown eyes and a quick, ready wit. He has lived in these ruins, eating garbage from the dump next door, pissing in the corners, defecating in the rubble outside the crumbling walls, for almost two years, and it is he who maintains the tenuous order of the place, throwing someone out if they become absolutely too disruptive and delivering the final verdict on who will be allowed to stay. His standards are not very high, but the others defer to him.

The residents come and go during the day. They circulate through the market to scuffle for a *cordoba,* and when they manage to get one, they buy a bottle of glue at a cobbler's stand and head back to the ruins. Inside the squat, there is a steady to and fro of teenagers with their adult heads and stunted bodies in various states of zombie-dom. They shuffle about or sit on the slick floor with their backs against a brick wall, staring out of vacant eyes, a baby food bottle held over nose and mouth, grime black under their fingernails or in the grooves of the skin beneath a nail gnawed away. When the glue wears off, and hunger becomes too insistent to ignore, a kid will go to the lot next door where a steady stream of wooden-wheeled carts arrive. Heaped high with garbage, pulled by burros, the carts are driven by people who make their livings circulating through the streets of the market all day, picking up refuse and hauling it away. Given the level of poverty and hunger in the streets of Managua, it is not hard to figure out that the food that makes its way to the garbage dump in the *Mercado Oriental,* food that has been discarded for being inedible, is, in fact, pretty nearly that. Nevertheless, the kids usually keep a little trash fire burning on the floor in a corner of the abandoned building, with a pot balanced precariously over it on a blackened piece of grill. Whatever they have brought back from the dump goes into the simmering stew in the pot.

People call the *huele pegas* violent and crazy, *el toro* told us, but it's just because some people don't have any better sense than to mess with someone high on glue. The glue makes you unpredictable, there's no telling how someone will react. As long as they're left alone, however, the addicts will not cause trouble. Life here is tough for these homeless youngsters, he said. The police beat them, the merchants hate them for their thieving ways, and adult lowlifes are always trying to beat, rob, or rape them. Such are the human perils of a life lived in the asshole of the *Mercado Oriental*. Also to be reckoned with, he added, are a *sin numero de bichos*, bugs and vermin without end, including rats, lice, and, of course, cockroaches. The roaches, too, hunt endlessly through the discarded garbage for something to eat, and he said he had bitten into more than one by accident.

. . . •

I haven't said how, sitting there motionless, I still hadn't stopped looking with deep disgust, yes, still with disgust, at the yellowed white mass on top of the cockroach's grayness. And I knew that as long as I had that disgust the world would evade me and I would evade myself. I knew that the basic error in living was finding cockroaches disgusting. Finding disgust in the thought of kissing a leper was my missing the primary life inside me...for disgust contradicts me, contradicts my matter in me....

For redemption must be the thing itself. And redemption in the thing itself would be my putting into my own mouth the white paste from the cockroach.

THE PASSION ACCORDING TO G.H.
by Clarice Lispector
translated by Ronald W. Sousa

• • • .

This novel is a prolonged meditation on a single cockroach, crushed and held in the door of a closet, with its fat body seeping out. A Ukranian-born woman who lived in Brazil and died in 1977 at the age of fifty-six, Lispector is described in the translator's forward as "a literary cause célèbre in her adopted

Brazil but is viewed in France, because of the very same texts, as an important contemporary philosopher dealing with the relationships between language and human (especially female) subjecthood. . . ."

. . • ●

There are not many places where cockroaches are intentionally eaten, nothing like the frequency with which people around the globe devour, say, larvae, grubs, or grasshoppers. Of course, grasshoppers fatten up on plant life, a much more palatable diet than garbage as far as humans are concerned. And, they don't smell bad. Roth and Willis report a theory that the disagreeable odor that is characteristic of many roach species is a survival tool to discourage people from eating them (Roth and Willis, 1957). They cite a 1932 report from southern Thailand that in certain districts Laotians ate roaches, but in most districts they did not because they smelled bad, although in all districts Laotians fried and ate oothecae, and children hunted for the egg cases.

The two researchers also note half a dozen other reliable reports from around the world about the ways in which people eat cockroaches. People on the island of Formosa were reported, in 1924, to have a recipe that called for removing the head and entrails, putting salt in the body cavity, and frying it. Roth and Willis also included some instances of reported roach eating that they viewed with a skeptical eye. For example, in 1905, in typical French anti-Anglo spirit, a Frenchman named Coupin put together a book about bizarre practices around the world, *LES BIZARRERIES DES RACES HUMAINES,* in which he included a recipe he claimed the English and Irish used to prepare roaches: "Simmer them in vinegar all morning then dry them in the sun. Discard their heads and intestines then boil them with butter, pepper, and salt to make a paste which is spread on buttered bread." There is no other evidence that people in the United Kingdom were eating roaches in those years, in place of

Marmite or other favorite English bread spreads, although there is a report that the feces of *Periplaneta americana* were being used in British homeopathic medicine. Henri Charriere, Papillon, is said to have caught and eaten cockroaches while a prisoner on Devil's Island in order to survive. Starvation is probably about the only reason most people would consider eating cockroaches.

While folks may not be enthusiastic about the idea of eating cockroaches, the bugs seem to have no problem eating humans. There are numerous reported incidents of cockroaches dining on human flesh from both dead and living bodies. Cockroaches are rip-and-tear eaters, they have no teeth or tongue, but they do have strong jaws and will soon leave a wound where they have fed. If a dead body is left undiscovered for a while in any cockroach-infested house, the roaches may do some serious damage to its flesh. One of the most unpleasant such reports comes from an article published in a 1997 issue of THE AMERICAN JOURNAL OF FORENSIC MEDICINE AND PATHOLOGY, entitled, "Cockroach: The Omnivorous Scavenger," and subtitled, "Potential Misinterpretation of Postmortem Injuries." It discusses three cases in which infants died suddenly, and their bodies were first discovered, and gnawed on, by cockroaches. When someone found and inspected the corpses, professionals misdiagnosed the bitten flesh as evidence of child abuse. In each case, before the sad truth was finally revealed, grieving parents were interrogated and implicated in their own child's death. The article is accompanied by black-and-white photographs of the dead babies and their wounds, as well as a word of caution to forensic professionals that they carefully assess such injuries to avoid misinterpretation (Denic, 1997).

THE AMERICAN JOURNAL OF FORENSIC MEDICINE AND PATHOLOGY is one of the most riveting periodicals in any medical library. It is a compendium of the myriad ways in which death can catch us unaware, full of reports about how to read the signs death leaves behind. The magazine is filled with case histories, and each details how death came, how it struck which organ, and with what degree of ferocity. Many of the articles have photos, and as shocking as they can be, it is the case histories, themselves, their cold and clinical laying out in words, that astonish the reader. Each death is as different as we are different, each a story to be told, a case to be reported.

As we all know, history repeats itself. Law enforcement officials were being cautioned fifty years ago to carefully assess wounds because of the potential for cockroach damage to a body. Roth and Willis cite one H. Hartnack who wrote in 1939 of corpses being found and thought to be the victims of crimes because of the skin defects that had been caused by cockroaches gnawing on their flesh. In fact, cockroaches and flies affect crime sites in Queens, New York, to such a degree that the police department's homicide division has a detective who is also a forensic entomologist, of which there are only a half dozen or so in the U.S.

A human does not have to be dead to be eaten by a cockroach. In the early 1950s, there was a brief flurry of debate among roach researchers as to whether or not cockroaches bite living people. They do. Roth and Willis addressed the question with their usual thoroughness and documented at least eighteen reports of cockroaches biting people. In many cases, the roaches were found to gnaw on extremities (Roth and Willis, 1957). A report in the December 8, 1870, issue of NATURE MAGAZINE, written by Arthur Nicols said the following: "In some ships infested with these insects, sailors frequently complain of having their toe and finger nails, and the hard parts of the soles of their feet and palms of their hands, nibbled by them. The men have exhibited to me their nails, and skin, which had the appearance of having been attacked. I can vouch for the following, as I was the unhappy subject of it. On returning from a shooting excursion in salt swamps in tropical Australia, with my feet blistered and sodden, I was put to sleep in a room swarming with cockroaches (the small species). The night was intensely hot, and my feet were exposed. I had slept soundly for some hours, when an intolerable itching and irritation about my feet awoke me. I felt these objectionable beasts running over and gnawing at my feet. On striking a light, I found they had attacked the skin, and entirely eaten it away from a large blister, leaving a raw place as large as a shilling. I slept again, and in the morning found they had completed the work, and established a painful sore. The whole of the hard skin on the heel was also eaten down to the pink flesh. The nails were not attacked. I have now, at a distance of four years time, bluish scars on the skin."

Sometimes it's not the flesh that the cockroach is after, but what's on the flesh. They are particularly fond of the traces of dried milk around a baby's mouth, and there are numerous reports of mothers who have surprised cockroaches feeding around the mouths of sleeping infants and children. This may have been what precipitated a tragedy that happened in an Atlanta, Georgia, public housing project in 1997. An eight-month-old girl, living with her mother, choked to death on a roach that entered her mouth while she was sleeping. While less extreme, some people have allergic reactions to cockroach saliva. Cases have been reported of children developing rashes around their mouths after suffering the depredations of cockroaches.

As early as 1747, Englishman Mark Catesby in his report entitled NATURAL HISTORY OF CAROLINA, FLORIDA AND THE BAHAMAS, wrote about the cockroach: "These are very troublesome and destructive vermin, and are so numerous and voracious that it is impossible to keep victuals of any kind from being devoured without close covering. They are flat and so thin that few chests or boxes can exclude them. They eat not only leather, parchment, and woolen, but linen and paper....It is at night they commit their depredations, and bite people in their beds, especially children's fingers that are greasy."

More than one hundred and fifty years later, in 1902, from another far-off, unexplored corner of the Americas, the naturalist Herbert Smith wrote to C. L. Marlatt, an entomologist at the U.S. Department of Agriculture: "...at Corumba, on the upper Paraguay [in Brazil], I came across the cockroach in a new role. In the house where we were staying there were nearly a dozen children, and every one of them had their eyelashes more or less eaten off by cockroaches—a large brown species, one of the commonest kind throughout Brazil. The eyelashes were bitten off irregularly, in some cases quite close to the lid. Like most Brazilians, these children had very long, black eyelashes, and their appearance thus defaced was odd enough. The trouble was confined to children, I suppose because they are heavy sleepers and do not disturb the insects at work. My wife and I sometimes brushed cockroaches from our faces at night, but thought nothing more of the matter. The roaches also bite off bits of toe nails."

People shouldn't take it personally that roaches might be inclined to nibble on some part of them, because cockroaches will, after all, eat damn near anything. Cockroaches take their nutritional value where they can find it, and they have been known to find it in some odd places. Paper will do just fine. As early as 1676, in England, minutes from a meeting of the St. George Council read: "Upon ye motion of ye Secretary to ye Governor & Councell that there was great need of a chest in ye office to secure ye Records ffrom the Cackaroches, wch did eate & deface papers. Ordered that ye Sherriffe cause a Chest to be made to secure ye records."

They are also known to prize the paste and glue that is used to bind paper together into books. For this reason, before the age of the computer, the dilemma of the St. George Council was widespread, and they frequently bedeviled paper-dependent bureaucracies. In 1902, C.L. Marlatt confirmed from Washington, D.C. that cockroaches thrive on big government. He reported instances of American cockroaches eating both cloth and leather bound books at the Treasury Department, and German roaches infesting his own Department of Agriculture: "The injuries effected [sic] by it to cloth-bound reports have been the source of very considerable annoyance at the Department of

Agriculture and in the large libraries of eastern towns and colleges. . . .In attempting to eradicate roaches from the Department storerooms where cloth-bound books are kept various paste mixtures containing arsenic were tried, but the roaches invariably refused to feed on them in the least."

The alimentary canal and digestive system of a cockroach is a fairly straight shot from mouth to anus, passing through three distinct parts—fore-gut, mid-gut, and hind-gut—with the whole system enclosed and protected by the fat body. When a roach eats, it raises its head to position its mouth vertically over the food, because at rest its mouthparts are tucked under, against its body. There is a crop, where the digestive process begins with digestive liquids breaking down the food. Behind the crop is the gizzard in the fore-gut which has tiny toothlike ridges that further help to break the food down. The food is then passed to the mid-gut where most of the nutrition is absorbed from it, and what is left moves on to the hind-gut through the Malpighian tubes and eventually out the anus. Experiments have been done using carmine dyes to determine how rapidly the excretory process functions in American cockroaches, and it was found that both temperature variations and pesticides affected excretion (Patton, 1959). When the insect eats, food reaches the crop after thirty minutes, but if it is a full meal, three days may pass before the crop is completely empty again. Roaches, like most beings, do have their food preferences. A study was done in which a variety of foods was laid out each night, and statistics gathered about which foods were chosen first by Oriental cockroaches. It turned out that they preferred sugary cinnamon buns above all else, after which came white bread, boiled potato, and sliced banana (Rau, 1945).

Of course, cockroaches both eat and are eaten. Theirs is hardly a place of privilege on the food chain. While few humans eat them, the roach has both external and internal predators and parasites. There are centipedes that have a primary diet of cockroaches. Mantises, ants, and scorpions will eat

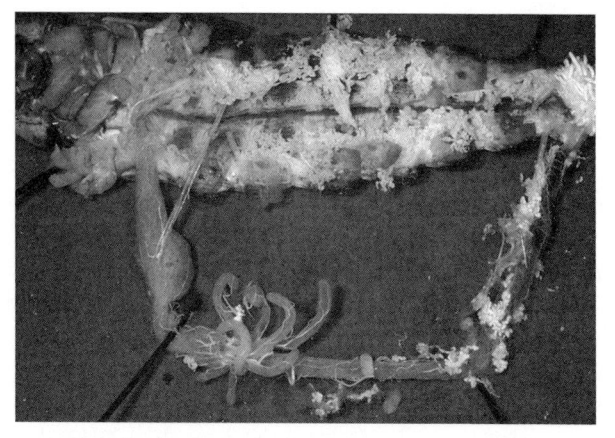

Figure 14. The digestive tract (from left to right, crop, gastric caeca, malpighian tubules, midgut, hindgut). © Betty Faber.

Figure 15. Cockroaches feeding. © Betty Faber.

them, as will a variety of larger animals including toads, frogs, possums, hedgehogs, armadillos, mongooses, monkeys, lizards, spiders, mice, cats, and birds. (A Spanish proverb says, "The cockroach is always wrong when arguing with a chicken.")

There are a number of species of small wasps that lay their young in cockroach oothecae, and when the wasp larvae hatch they feed on the cockroach eggs. One such wasp, *Aprostocetus hagenowii,* is the size of a fruit fly and only lives a week as an adult, going through its entire life cycle with little effect on its environment other than to destroy cockroach eggs. There are other species of wasps that prefer their roaches in the nymphal stage, and still others that prefer to dine on adults. They sting a roach to paralyze or stun it, then drag it back to their nests by one of its antennae, and lay an egg on it. When the larva hatches, the paralyzed roach serves as its first meal.

In addition, the gut and fat body of most roaches are home to a variety of "bacteroids," which help the cockroach manufacture vitamins and keep it healthy (Roth and Willis, 1960). Roaches harbor many bacteria and viruses, and they can be colonized by cockroach mites, some of which do the roach no harm. These mites exist in a relationship to roaches that parallels the one roaches have with us. They live on the roach and eat bits of food the roach leaves behind. Other mites seem to feed directly on the insect itself, as was demonstrated by giving food marked radioactively to an American cockroach from the body of which radioactive mites were later recovered (Cunliffe, 1952).

.   •   •   ●

Roaches are nocturnal and pass their days sleeping, or languidly lounging around their harborages. When you see roaches out and about during the day, it's a bad sign, because it means that they have been crowded out of their shelter and into the light. It is a sign of a serious infestation. Not all roaches come out at night, but all the periodomestic ones do. There are species of roaches in the wild around the world that have a diur-

Figure 16. Cave roaches. © Betty Faber.

nal circadian rhythm, i.e., they do their business in the daylight. Little or nothing is known about the activities of those in the wild. Thousands of species are yet to be collected and identified, much less studied. The lifestyles of domestic roaches have been exhaustively observed, of course, in the hopes that more knowledge will make it easier to eliminate them.

Normally, a domestic cockroach carries on its life in a world parallel to ours, behind wallpaper, hidden in the cracks of kitchen cabinets, under the refrigerator, or near the toilet where the pipes pass through the bathroom floor. It comes out in the dark to work for its daily bread, to find food, water, and a mate. The instances are few when its path will cross with that of humans, and whole generations may be born, live, and die without ever being seen by a human eye, even though their worlds occupy the same space and they may be only inches apart as they move through their respective wakings and sleepings.

Neurobiologists have spent an increasing amount of time and resources studying circadian rhythms over the last fifty years. These are the doings and restings, the eating and sleep-ings of living creatures as regulated by a "biological clock," now believed to be part of the make-up of most sentient creatures. The cockroach has proven a favorite for researchers in this area. In1955, an entomologist named Janet Harker in Cambridge, England, was painting the heads of roaches with black paint, blinding them, to see if they kept their nocturnal rhythm, even if they could not see to distinguish night from day. She found that they gradually changed their activity schedule, becoming active during some daylight hours, and staying inertly harbored during more evening hours than the unpainted control cockroaches. However, if she implanted the suboesophageal ganglia from a normal cockroach's abdomen into the abdomen of one whose eyes were painted shut, the blind insects regained their natural circadian rhythms (Harker, 1955). From this, she deduced that the ganglia exercised a neuro-endocrinological control of a roach's schedule.

A dozen years later, in 1967, another Cambridge researcher, John Brady, began his article in the JOURNAL OF EXPERIMENTAL BIOLOGY by discussing Harker's work: "Only a very few physiological clocks have been located accurately in any organisms; two of them in invertebrates are well documented." Nevertheless, he went on to write that he had little success in replicating Harker's experiment, and posited the existence of a control mechanism—"an electrical pacemaker"—for circadian rhythm that would be found in the brain (Brady, 1967). These experiments, and the many that followed, measured a cock-roach's activity over twenty-four-hour cycles.

The search for that electrical pacemaker in cockroaches has occupied a dozen or so biologists around the world for the past quarter of a century. Terry Page, at Vanderbilt University, is one of them. He has been conducting his search for more than twenty years, funded primarily by the National Science Foundation. He offered to show me a cockroach's brain. I fol-lowed him into the small room next to his lab at Vanderbilt, where he kept his colony of *Periplaneta americana*, the long, dark roach that most people genteely refer to as a water bug. The windowless room had what I had come to know as the smell of an infestation of cockroaches: a disagreeable tangy mustiness like a lot of wet cardboard had been left piled over some small dead animal in an unventilated corner. He reached into a plastic bin and took out a long, shiny American roach between thumb and forefinger. We walked next door. He put the roach in a covered plastic dish with a tube entering one side, through which he pumped a little carbon dioxide. CO_2 anesthetizes cockroaches, and this one ceased moving its antennae almost immediately. Page had another small, round plastic dish with a V-shaped notch in it, into which he positioned the unconcsious roach's neck, so that its head extruded through to the other side of the plastic, while the rest of the body was in the dish behind. All of this was clamped in a holder beneath a dissecting microscope. The roach's head looked like a little brown knob to the naked eye, but under the microscope it became the huge bug

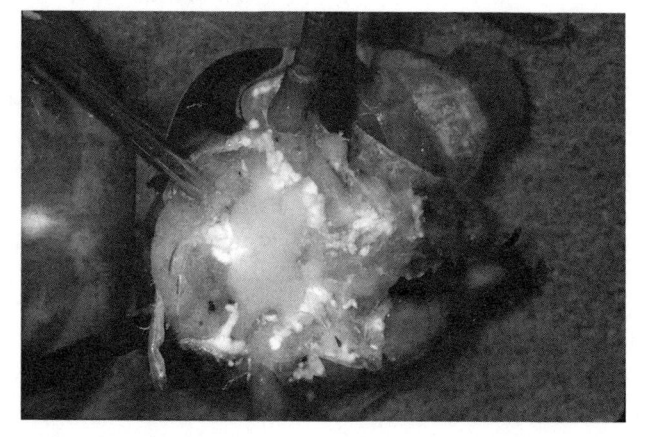

Figure 17. A cockroach's brain (supraesophageal ganglion).
© Betty Faber.

face of our horror-movie dreams with its multi-faceted eyes and insect mouth.

"It's an interesting thing, but after you've done this as many times as I have you realize that each cockroach has its own individual face," Page told me. "Each one is slightly different."

So saying, he used a single-edge razor blade and tweezers to make a slice down the middle of the roach's forehead and peel the flaps of skin back from it. He taped them down on either side with a little square of white tape. A pearl-gray glob of brain was exposed. Page explained that after years of experiments, researchers found that when they cut the nerves between the eyes and the brain, the insect's circadian rhythms remained the same, but if they cut the connection between the optic lobe and the rest of the brain, the biological clock was disrupted.

"That led people, including myself, to look for the part of the optic lobe it was in. We made smaller and smaller lesions, destroying little bits of the optic lobe and the result was consistent that there was a discrete area necessary to maintain this activity. Of course, with lesioning you can never be sure what you have done. You may have destroyed something that wasn't the biological clock and changed the activity that way."

How do you "lesion" a cockroach's brain?

"You take what is, essentially, a pin, sharpen it electrolytically, and then you put it in a device called a micromanipulator, which allows you to position it very precisely," Page told me. "You stick the pin into the brain and then you pass a little bit of current through it. This heats up the tip of the pin and you burn out a very small portion of the brain. A little sphere of 50 or 100 microns in diameter depending on how much current you've used."

He untaped the skin flaps and rejoined them at the forehead with low-melting-point wax. The roach would wake up in five or ten minutes as good as new, he told me, as we continued our conversation in his office. "My research is, obviously, not directly applicable to human beings, but the kinds of things I'm interested in are fundamental processes, and it's reasonable to assume they are cross-phylogenetic. When I started working on cockroaches one of the issues was just how localizable is the biological clock? Is it actually a very discrete localized structure, or does it involve diffuse interconnections among lots of cells with various cells being able to carry out the function?

"At the time, that wasn't clear for any organism and so it seemed a question that could be of general interest. These things have applications to such human problems as jet lag, for instance. While our observations with cockroaches are not directly applicable to human beings, they provide a sort of a conceptual framework as to how things might work. That makes it easier and helps guide experiments on more complicated organisms. If you do these experiments on cockroaches and snails and frogs and fish, and you always get the same answer, then you begin to believe you're uncovering an organizational principle."

He leaned back in his chair as he explained this to me. A tall, lean, healthy-looking man, he cupped his chin in his palm and pulled down the edges of his lower lip with thumb and forefinger in a thoughtful gesture. The same fingers he had used to lift the roach from the plastic bin.

· · • ●

Given a choice, a cockroach from one of the species that lives with humans will pass its days crammed into as small a space as possible with as many fellow roaches as possible. There is no such thing as a claustrophobic cockroach. The more hemmed in they are, the tighter the space into which they are squeezed, the more secure they feel. Given a choice, a cockroach will be touched on all sides. This preference for contact is called positive thigmotaxis by scientists and the German cockroach is an example par excellence. It will seldom choose to rest in a space that's more than ½-inch high, or less than ¹⁄₁₆-inch low (Berthold, 1967). One-quarter of an inch is just about right to feel boxed in, preferably crowded by other roaches on all sides.

Researchers have known for a long time that roaches reared in isolation take longer to go through their molts and reach adulthood (Pettit, 1940). It is reasonable to posit that a cockroach may define home as anywhere there are a number of other cockroaches living comfortably. This does not mean that a community of cockroaches is always a peaceful place where an individual derives continuous satisfaction from the presence of others. Cockroaches, like so many other species (including our own), have male aggression rituals. They have their own inventory of aggressive behaviors, a scale of conflict that begins with threatening postures. Beyond that they graduate to antennae lashing—a form of which is also present in male/female encounters to determine if a female is sexually receptive—and biting. Sex and territory seem to be the primary motivations for fighting between male cockroaches: these clashes never end in death, but always in the retreat of one fighter (Bell and Sams, 1973).

A domestic cockroach's concept of territory remains unclear. The question of how much a roach will wander under normal living conditions has yet to be definitively answered. Experiments done in the 1950s in sewer lines, introducing marked cockroaches under manholes and setting traps in hopes of recovering them, came up with differing results. Some seemed to indicate that roaches traveled as much as three hundred yards on a regular basis, and other studies suggested season and population density were determining factors (Cornwell, 1968).

. . . •

Since my return to Paris I have been saddened as never before by the anonymous crowd I see from my windows engulfing itself in the Metro or pouring out of the Metro at fixed hours. Truly, that isn't a life. It isn't human. It must come to a stop. It's slavery....

French writer Blaise Cendrars,
from a 1952 interview in *PARIS REVIEW*

• • • .

In order for scientists to study the movements of roaches, they must mark them, and then trap them. In the process, they have experimented with many different kinds of bait in traps, but most have determined that stale white bread moistened with warm, slightly soured beer is the most reliable and effective. This is typically placed at the bottom of a small jar—a baby food jar, say—around the interior rim of which a petroleum jellylike Vaseline has been applied. The cockroach can climb in from the outside, but can't climb back out once it is inside. This is a simple kind of trap that can be used by anybody in their own kitchen to determine whether they have cockroaches, as well as by entomologists with Ph.D.s working in laboratories with big budgets.

These were the kind of traps used to measure German cockroach movement inside a group of apartments in a public housing project in Indianapolis in 1978 (Owens, 1982). The scientists involved were using what is called the "mark and recapture" technique. Roaches were marked with tiny dabs of Testor's model airplane paint, and traps were placed in eight apartments. The housing project was described as having a "chronic and generally severe" roach infestation. The results seemed to indicate that the German cockroaches involved in the experiment did a fair bit of traveling to and from various apartments. The roaches used the building's interconnected plumbing system like a highway on which to travel, and the experiment concluded that as much as 30 percent of the population moved between apartments over the course of a week. Other experiments have shown, however, that while roaches may move between apartments, they are not inclined to move between buildings, unless circumstances force them to do so. As long as the living is good, cockroaches seem content to stay pretty close to wherever they first called home.

In addition to how a cockroach perceives territory, another related, unresolved question is why a cockroach elects to go where it goes or avoids the places it chooses not to go. Here, again, pheromones appear to play a role. Researchers at the University of Florida have shown that young roaches do not have to discover their own routes to find food and water, but instead follow the fecal trails laid down by other roaches from the same harborage.

One thing seems certain: there is more at work than simple random rambling when a roach is on the move. As early as 1884, experiments done in Germany proved that cockroaches head for dark places during the day and are most active at night. Less than thirty years later, it was also shown that if given a choice of going to a dark area during the day and getting shocked, or huddling up in the light, the roaches soon learned to remain in the light and avoid the dark altogether (Szymanski, 1912). In that same year, C.H. Turner, an African-American biologist at Sumner Teachers College in St. Louis, confirmed these results, and shortly thereafter he successfully taught cockroaches to run a maze, by building it over water so the roaches wouldn't just wander away from the test (Turner, 1912). Over time, he found that the roaches were able to run the maze in shorter and shorter times, which suggested that they had learned the route. Turner had to build the equipment for these tests himself, as the budgets for black biologists at teachers' colleges were extremely limited. Turner left a mark with his research, and his results are still referred to today.

In the 1960s, a series of remarkable experiments were designed using electric shock avoidance to train a cockroach suspended over water not to dip its leg down. If the end of the leg touched the water, the roach was shocked. This lesson was quickly learned by the insect, so that the chosen leg would be held under the body and not straightened when the roach was suspended. The remarkable thing was that this behavior was exhibited even after a cockroach had been decapitated (Horridge, 1962), which indicated that learning was taking place at more than one site on the body. Researchers considered the possibility that the gathering of nerve fibers close to the cockroach's abdomen might function as a second brain, retaining information in a kind of back-up system.

·   ·   •   ●

man is a queer looking gink
who uses what brains he has
to get himself into trouble with
and then blames it on the fates
the only invention man ever made
which we insects do not have
is money and he gives up
everything else to get money
and then discovers that it is not worth
what he gave up to get it
in his envy he invents
insect exterminators
but in time every city he builds
is eaten down by insects
what i ask you is babylon now
it is the habitation of fleas
also nineveh and tyre
humanitys culture consists
in sitting down in circles
and passing the word around
about how darned smart humanity is

from "quote and only man is vile quote"
in THE LIVES AND TIMES OF ARCHY & MEHITABEL
by Don Marquis

●   ●   ·   ·

Don Marquis's cockroach character, Archy, made his first appearance in 1916 in the "Sun Dial" column of the NEW YORK SUN, and became, perhaps, the best-known cockroach of the twentieth century to North Americans. The Archy poems, which also include the voice of the bug's friend Mehitabel the cat, are great poetry, with a humor as sharp as the grave, readable by both old and young. Column writing for most journalists is a pasture to be put out to, a thankless daily slog through the same mental swamp to mine a certain standard number of words that push the same old buttons day in and day out. Marquis, on the other hand, by using a cockroach that was a poet (Archy wrote by banging his head on the keys of his boss's typewriter and could not use the shift key, so neither capital letters nor punctuation ever appeared in his poetry) turned his column into a wonderful, corrosively funny place for readers to stop in each day.

The only other candidate for the century's best-known written roach might be the bug in Franz Kafka's short story, "The Metamorphosis," about a young man named Gregor Samsa, who wakes one morning to find himself turned into a giant insect, flat on his back, legs waving in the air. Kafka always denied that Samsa was, specifically, a cockroach, saying that to identify the species of bug into which his protagonist suddenly turned would be to weaken the story. When his publisher wanted to use an illustration from "The Metamorphosis" on the cover, Kafka insisted that it not show the insect, itself. He wanted each reader to imagine his or her own most repulsive insect. That repulsive insect turned out to be a cockroach for so many people that despite all of Kafka's cautions against specificity, many people now think of the story as the one in which a man wakes up as a roach.

"The Metamorphosis" somehow strikes a deep and resonating chord in readers all over the world. Perhaps they feel that the bodies they inhabit are more temporary and provisional than they might like, or that their families' love for them is more conditional than it might seem. Truly few are those who would stand by us if we woke up one morning in a beastly bug's body. One Nobel prize-winning writer who said the story was instrumental in initiating his writing life is Gabriel García Marquez, whose ONE HUNDRED YEARS OF SOLITUDE is widely considered the best example of what is called magic realism, as well as one of the century's best novels. In a 1988 interview in EL PAÍS, the Spanish newspaper, which I've translated, he remembered with great clarity the moment Kafka's story opened up new creative vistas for him:

"Kafka's 'Metamorphosis' was a revelation . . . it was in 1947 . . . I was nineteen years old . . . I remember the first phrase, it goes exactly like this: 'Gregor Samsa awoke one morning, after a restless dream, to find himself in bed transformed into a monstrous insect'! . . . Damn! . . . When I read this I said: But this doesn't count! . . . Nobody told me you could do this! But if you can do this . . . then I can do it! . . . Damn! . . . This is how my grandfather told stories. . . . The most unusual things, with the greatest naturalness."

. . • •

For all my wasted worry, the hissers and I passed across borders as easily as a snake going down a hole. In no time the Madagascars were installed in a corner of my study, inside a high box made of transparent plastic. I put a red gel over a flashlight at night and watched them. At one point, I actually saw a female with an ootheca extruding and remembered Lou Roth's pregnancy diagnosis. In fact, it was less than three weeks after that sighting when I saw first instars clumped up together in a corner, and shortly thereafter I counted thirteen juveniles, in about their third instar, tearing around the walls of the terrarium. The small ones were fast as hell when they moved, zipping around. They were kind of cute, miniature Madagascar roaches, and if they were not moving, they were huddled up all together in a clump, reminding me of how catfish fry will invariably bunch up together at one place underwater near a pond bank, forming a wriggling mass of thousands of tiny catfish. The thirteen Madagascar nymphs all gathered in a wrinkle of the paper on the bottom of the terrarium in one dark twitching mass.

Eventually, though, it began to look a little crowded in there, so I called an entomologist I knew at one of Spain's most important government research institutions, who is working on developing an environmentally friendly insecticide that would interrupt the growth of the egg in the oocyte of German cockroaches. Most scientific investigators in Spain have to rely on public moneys to fund their work. It is a country without a tradition of cooperation between capitalism and science, but my acquaintance had a research grant from a private-sector company—not surprisingly, an insecticide producer. He was the person who first told me about Lou Roth and his work with cockroaches, so it seemed only right that I should return the favor. I called and inquired if he would like some Madagascar hissing cockroaches to keep in his lab.

"Live ones?" he asked, hope in his voice. When I answered affirmatively he told me to come by whenever I wanted.

"I'll be right over," I said. I put the terrarium on the Peugeot's back seat and drove to the Barcelona laboratory. A short, bearded, balding man, he and his assistant, a tall, unsmiling woman, were waiting there in their long white lab coats, and they had a cage prepared. I put the terrarium on the lab bench beside it and stood back as they opened it and began dividing the colony. At one point, as she was transferring one of the Madagascars, it surprised her with a loud, sharp hiss. Taking advantage of the slight, reflexive opening of her hand in response to the fierce noise, the roach jumped out of her grip. All six legs were churning, and within a split second it was off the countertop. Fortunately, it fell into a large trash barrel beneath the edge of the bench. Both the professor and his assistant looked into the barrel, as did I. Full of crumpled paper towels, there was no roach that any of us could see. They glanced at each other, and I thought, this must be the look that nuclear power plant operators exchange on the verge of a major problem. Both of them seemed to briefly imagine Barcelona ten years later, infested with gigantic hissing cockroaches, a real possibility if the roach somewhere in that barrel happened to be a pregnant female. The professor began gingerly lifting out trash until he spotted the roach crouched in a piece of rumpled paper towel. He swooped down on it and carried it, hissing, to its new cage. It was a big male.

"It's a beauty," he said, real admiration in his voice.

a bullet don't have nobody's name on it

IN THE LATE 1970S, I worked for almost two years as a criminal investigator on felony cases for the public defender's office in Portland, Oregon. My job was to read all the notes and the discovery material for a given case and try to thwart the prosecution's presentation. I looked for new witnesses, previously undiscovered facts, a weak link in the D.A.'s version of what happened—anything to produce reasonable doubt in a jury, or, better yet, to convince the D.A. to cut a pretrial deal. Almost without exception our clients were guilty, so we tried to get them the best deal we could fashion by creating an opportunity for a plea and a reduced sentence.

My first homicide case took me to the housing projects of North Portland, where a woman who weighed in at an easy two hundred pounds opened the door to my knock. The fact that she consented to do so was, in itself, an achievement when measured on the scales used by most criminal defense investigators in Portland, many of whom were ex-cops who still had the appearance of the Law. They dressed in coats and ties, carried briefcases, and having used authority as their modus operandi during their careers in law enforcement, they were hard-pressed to project any sort of sympathy, whatsoever, toward criminals or people living their lives in dubious circumstances. For the most part, the people they wanted to talk with would not respond to their knock, let alone talk to them. At most, they cracked their front doors the width of a safety chain to give monosyllabic answers to questions. I dressed as I always did with an open collar, no tie, and no briefcase. I kept my notebook in my back pocket and looked a lot less like trouble coming. People were more inclined to allow me inside and to sit down and speak to me. Of course, most of the time they still tried not to tell me anything useful, because they all knew how the law-and-order game worked. They were well aware that if they gave me something that could help my client, they could expect to be subpoenaed to say it in front of a judge and jury. Courtrooms were somewhere most of the people to whom I talked tried to appear in as little as possible.

"I didn't see a thing," she told me, of the night in question, when a party she was attending at a nearby apartment was broken up by the shooting death of which our client was accused. "I didn't see a thing. We were sitting at a little ole table drinking whiskey and laughing, me and three of my girlfriends. There were some guys in the other room, and when I heard the first shots I got under that table as fast as I could move all this flesh, honey. A bullet don't have nobody's name on it.

"I don't know anything, 'cause I didn't see anything. I was keeping my head down. A bullet don't have nobody's name on it," she repeated.

A little boy, who couldn't have been more than six, wandered, barefoot, into the living room, a square, cement cinder block apartment in the projects, with a concrete floor. "I can't breathe, Mama," he whimpered in a tight voice, one small hand clutching his chest.

"Come here, baby," she said, taking an inhaler off the low, rectangular coffee table in front of her, where it sat beside her ashtray and an open pack of cigarettes. The boy came over and put his mouth around the inhaler in a practiced movement, while she pushed the plunger down and gave him a dose. "Go on back in there and watch cartoons, now," she said to him. "I'll be done in a minute."

I stood up to leave. "He's got the bad asthma," she told me. "Seems like we spend half our time in the emergency room."

Something in my own chest tightened in sympathy. My childhood and adolescent asthma began to go away, oddly enough, following the first time I had sex. But I have a memory as clear as water of the panic engendered by a tightly constricted chest. I spent days at home from school with asthma in my Nashville childhood. Propped up on pillows in bed—it was easier to breathe sitting up—I struggled to complete that most basic of human functions, taking air in. Wheezing, I listened all day to the rock 'n' roll station of those years, WKDA, play Elvis Presley and Little Richard. I remember how an inhaler would stop a mild attack and make my heart race. On the worst days, when even the inhaler did not loosen the knotted vessels of my lungs, the family doctor would make his house call and liberate me with a syringe of Adrenalin, which, when injected into an arm, would pry my chest open within seconds so the air could flood in. Oof, what a rush of compassion I suffer for the boy I was.

The little boy in the Portland housing project was a typical asthma sufferer, one of a rapidly growing number of poor children suffering from the disease in this country. This trend has alarmed both health care experts and social scientists. The total of both children and adults with asthma in the U.S.—10 to 12 million—has increased by about 60 percent over the past decade. There were over 500,000 hospitalizations for asthma in the U.S. in 1994, and more than a third of them were patients under eighteen. There were also some 2 million emergency room visits, and the total annual cost of treating asthma in the U.S. was more than $6.4 billion by 1990.

African American males who live in a large city are at almost three times greater risk of developing asthma than any other population group. Asthma is debilitating and terrifying, but rarely fatal. However, it can, and does, kill, and there are records of asthma deaths stretching back to the second century B.C., when Aretaeus of Cappadocia was the first writer to describe the disease, and noted that it could suffocate its victims. While deaths from asthma are still rare, they are increasing, and rising most dramatically among poor, younger males. In 1978, there were less than 2,000 deaths per year from asthma in the United States, a number that had jumped to over 4,500 by 1998. Black, inner-city males are most likely to die of it, and a full 21 percent of the asthma deaths in the age group of five to thirty-four in 1992 happened in Chicago and New York City. In the five boroughs of New York, 120,000 school-age children are estimated to suffer from asthma, and it is the number one reason for school absences.

Apart from his race, age, and city of residence, there were other factors in that North Portland boy's life that put him at risk for asthma. One was that his mother smoked, and he was exposed to a considerable amount of second-hand smoke. Another risk factor was that she likely had neither time for proper household maintenance, nor money to hire someone to do it, and the apartment probably had a substantial amount of dust.

Both cigarette smoke and dust are recognized as factors in the onset of asthma, but neither is the greatest allergen in an inner-city environment. That distinction belongs to cockroaches. For me, in the roachless home of those Nashville suburbs, it was irrelevant that in the past twenty years cockroaches have been identified as a leading cause of asthma, but for a boy growing up in a housing project, they may be why he has the disease, and they are almost certainly a contributing factor to the onset of his attacks.

A connection between cockroaches and asthma began to be established in the early 1960s, when it was discovered that skin tests revealed an allergy to cockroaches in some 28 percent of patients suffering from allergies (Bernton and Brown,

1967). Subsequent tests provided more data. "Among almost six hundred allergic patients belonging to four ethnic groups, routinely visiting seven hospitals in New York City, over 70 percent reacted positively to the cockroach allergen," wrote Cornwell in 1968. "Positive reactions were most marked among Puerto Ricans (59 percent), less marked among Negroes (47 percent) and Italians (17 percent) and least among Jews (5 percent); this is the same order as the severity of cockroach infestations (*B. germanica*) reported in the homes of these four groups in New York City."

Once the first studies were published linking cockroaches and asthma, the data began to come in from far and wide. Studies in a long list of countries, including Egypt, France, Switzerland, Japan, Portugal, New Zealand, Holland, Mexico, and Spain revealed the same thing: the presence of cockroaches can produce an asthma attack, or that of a closely related illness called rhinitis.

By 1993, the federal government was convinced enough to fund a federal study in eight U.S. urban areas, which looked at 1,528 children under age 10 with asthma who lived in the inner city. A large and careful study, which cost $17 million, it was published in the NEW ENGLAND JOURNAL OF MEDICINE in May 1997. The authors concluded that cockroaches were the single most probable environmental allergen to set off an asthma attack among inner-city children from eight U.S. cities. The specific chemical triggers appear to be proteins, according to Dr. David Rosenstreich of Albert Einstein Medical College in the Bronx, lead writer of the article.

"These are just ordinary proteins that are part of what makes up a cockroach," he told me. "There are at least six or seven allergens that I know of in cockroaches. Various parts of the cockroach are allergenic. There are substances in the saliva, feces, and in the blood, all of which are proteins that people are allergic to. When people are exposed to them in any form they will react to them.

"The usual form of exposure is when the roaches die or molt and become part of the dust, or their feces become part of the dust. People breathe that in and they get allergy problems. The most common of these allergies in the study is asthma. Cockroaches are a problem because there are so many of them and we're essentially locked in with them."

Of the inner-city children studied, almost 37 percent of them tested positive to cockroach allergen, the highest number for a single cause of allergic reaction. This group was followed by some 35 percent who were allergic to dust mites and 23 percent who reacted to cat allergen. Over half the children studied had "high levels" of cockroach allergen in dust collected from their bedrooms, meaning that cockroach feces, discarded exoskeletons, or dead roaches were present.

However, like other ills that plague our society and prematurely end lives, such as guns, drugs, or AIDS, asthmatic reaction to cockroach allergens is not something confined to the inner city. A 1994 study in Kentucky showed that while people living in the suburbs had a relatively low rate of allergy to cockroaches, those living in the inner city of Louisville and those living in rural areas or small towns had comparable allergic reactions, with the strongest reactions in both groups coming from children between the ages of seven and twelve (Garcia, 1994). These tests were done like all allergy tests—a cockroach extract was applied with a skin prick and if a "wheal" formed around the site, it was considered an allergic reaction.

Cockroach extract? Hey, somebody has to do it, and there are at least six commercial laboratories across the country that offer it for sale. Should you want to try it at home, the following recipe was kindly provided in an issue of the JOURNAL OF ALLERGY AND CLINICAL IMMUNOLOGY: "German cockroach (Blattella germanica) frass [cockroach debris containing body parts, fecal material, and egg cases] was obtained from cockroach-culture jars. Twenty-eight grams of frass was ground into a thick paste with a mortar and pestle and added to 140 ml of BBS [borate-buffered saline], pH 8.0. The material was stirred overnight at 4º C and then was centrifuged at 18,000 rpm for 1 hour. The supernatant was dialyzed extensively against BBS and ether extracted" (Pollart, 1991).

It has been well over a quarter century since the first papers were published indicating that cockroaches could cause an allergic reaction, but doctors who treat asthma have been slow to accept that cockroaches could be a primary culprit in the disease. For many years, it was assumed that the rise in asthma cases was connected to increased air pollution, but stricter environmental protection laws have reduced air pollution in many cities at the same time as the number of asthma cases have continued to climb.

"The inner-city study was so large, and so well done with so much data that it has pretty much led to a general acceptance that this is a problem," said Rosenstreich. "It may not be the only one, but it's significant."

Was the study worth $17 million? "On the one hand, the study's conclusions may seem obvious, but on the other hand they have motivated people to think about doing something about it," he said. "The government is now doing a whole big study focusing on cockroach eradication. It has motivated lots of people to do something about this problem by highlighting something that everyone was aware of but didn't realize how important it was relative to all the other things."

It may take people a while to link asthma and roaches in their minds, but for centuries there has been a general sentiment among the population that cockroaches must be vectors and spreaders of disease. Why else would people find them so repulsive, if not because they represent a real, not an imagined threat? But, the fact is, that while they certainly can serve as both vectors and disseminators of disease, they only do so infrequently. In fact, setting off asthma and allergy attacks represents the most general and widespread threat to human health from cockroaches yet to be identified.

This is not to minimize the negative effects cockroaches can have on our well-being. While the harmful and noxious effects of roaches were posited for a long time, it was only in 1957 that Louis Roth and Edwin Willis, working out of the U.S. Army lab in Natick, Massachusetts, published THE MEDICAL AND VETERINARY IMPORTANCE OF COCKROACHES, which systematically investigated the literature around the world regarding disease transmission by roaches, and grouped the results.

What they found was that the *Blattarians* are capable of transmitting disease to humans in just about all the forms to which we are vulnerable: virus, bacteria, fungi, and protozoa. Roth and Willis approached the question of the cockroach as transmitter of disease by collecting accounts of both those diseases that cockroaches were found to be carrying naturally, and those that they seemed capable of transmitting under laboratory conditions. Polio, for instance, was among the first category, in that German cockroaches living in the homes of people suffering from polio were found to be carrying the virus. In the second category, experiments also showed that American cockroaches fed an emulsion of rat brain with the polio virus in it had the virus in their feces for up to sixteen days after eating.

Fecal matter is one of the great vectors of disease. In reviewing the literature, Roth and Willis concluded that there was no doubt cockroaches regularly come into contact with human feces. They will eat them, and they frequently make their homes in sewers where they will have an opportunity to do so. In addition, in underdeveloped countries, outhouses are frequently the first choice of cockroaches as residences.

In a case that came to light after Roth and Willis published their monograph, cockroaches came under suspicion as carriers of the hepatitis virus in southern California (Tarshis, 1962). Between 1959 and 1961, the number of infectious hepatitis cases in Los Angeles County more than doubled—from 1,147 cases to 2,429. In addition, between 1956 and 1959, one housing project in the county had up to 39 percent of all the hepatitis cases. However, when the housing project was treated for cockroaches in 1960 and 1961, during the time that the overall number of hepatitis cases in the county was growing rapidly, hepatitis in the project fell to only 3.6 percent of the total, leading researchers to conclude that cockroaches might be one of the ways in which the disease was being spread.

About forty species of pathogenic bacteria have been found in cockroaches, including bubonic plague and leprosy (Hansen's disease) in areas where these diseases are common. Among the bacteria from which North Americans frequently suffer and that have been found in roaches is salmonella, which causes food poisoning and gastroenteritis. Nine different strains of salmonella have been found in American cockroaches taken from a variety of sites, including sewer manholes and private homes. In addition, there are a number of records of gastroenteritis outbreaks among patients in hospitals that ceased when stringent cockroach extermination methods were employed.

In fact, hospitals are one of the favorite residences of both German and American cockroaches, and it is in these institutions that roaches take on their most sinister aspects. In a hospital there is a high concentration of transmittable diseases, as well as a human population that is generally weakened and susceptible to them. Among the bacteria that has been found in, or on, roaches and their feces was an unidentified species of *Streptococcus*, associated with pus and found on the outer surfaces of nineteen cockroaches captured in a German hospital operating room; *Salmonella morbificans,* a cause of gasteroenteritis, from the feces and alimentary tracts of sixteen roaches captured in a hospital ward where there were cases of the disease; *Salmonella typhimurium,* from the gut of a roach caught in a ward where Salmonella infections were occurring; and *Shigella paradysenteriae,* which causes "summer diarrhea" in children, and which was found in a cockroach trapped in a hospital food cupboard.

Another favorite haunt of roaches that is cause for concern is the supermarket. Frequently, these businesses are closed and empty of humans at night with little to impede the free rambling roach and plenty to attract it. By the time the first employees arrive in the morning, long after the sun has risen and the roaches have returned to their harborages, there will be no way to determine what food, if any, has been contaminated by passing roaches. Scientists have taken electron microscope photos that show bacteria attached to cockroach legs, and it is not hard to imagine that food could be contaminated when a roach walked over it. In addition, professional exterminators agree that supermarket bags are one of the great contemporary transport methods for cockroaches, and they caution people to keep a weather eye out when putting away the groceries. Many a home has been infested by cockroaches brought back from the grocery store.

Another category of illnesses associated with cockroaches is that caused by protozoa, one-celled organisms that live in the gut and get their nourishment from food particles. Among these little bastards is the particularly unpleasant *Entamoeba histolytica,* responsible for amoebic dysentery. Both American and German cockroaches have been found carrying this protozoa. In Cairo, Egypt, researchers collected cockroaches in restaurants and found that 3 percent of them carried the protozoa for amoebic dysentery, while in Peru the number was 7 percent. In an experiment conducted in Venezuela, the feces of infected cockroaches were fed to young kittens that promptly developed amoebic dysentery, indicating that the amoeba was eminently transmittable.

Cockroaches have also been implicated in transmitting fungi. These can be particularly dangerous in hospital settings where types of fungi like aspergillus can appear in operating theaters and represent real threats to the health of patients, occasionally resulting in death. As early as 1908, aspergillus was isolated in the gut of cockroaches by French biologists.

The last category of illness in which roaches have been named as suspects involves those transmitted by helminths, which are parasitic worms, among whose members number such unpleasant little creatures as hookworms and tapeworms. Many kinds of helminths find cockroach guts to be magnificent places to grow up. People who have suffered roach infestations might be glad to see the bug, itself, being infested, but there is a possibility that the roaches could also transmit these worms to humans, via their feces. Roth and Willis gathered a number of disturbing reports, including one of 788 cockroaches examined in 1935 in Saint Petersburg (it was Leningrad then), in which roaches carrying pinworm eggs were found in a factory kitchen and a restaurant. Tapeworm eggs have been found occurring naturally in the American cockroach. Hookworm eggs and larvae were found thriving in American cockroaches in South Africa in 1929.

In addition to serving as vectors for disease, cockroaches also occasionally represent direct threats to the well-being of humans. They have been reported to have accidentally entered and become lodged in just about all the orifices of the human body. It is not unusual for emergency rooms in inner cities to treat people who have had a cockroach crawl into their ear and get stuck there, something that rapidly puts the unfortunate victim on the edge of madness. It is, according to all accounts, painful and horrifying, although a little mineral oil or lidocaine sprayed into the ear is usually enough to dislodge the intruder.

It is not a new problem. In 1749, an early Swedish settler in the New World related a similar experience to traveler and botanist Peter Kalm: "An old Swede . . . told me that he had in his younger years been once very much frightened on account of a cockroach which crept into his ear while he was asleep. He awoke suddenly, jumped out of bed, and felt that the insect, probably out of fear, was endeavoring with all its strength to get deeper. These attempts of the cockroach were so painful to him that he thought his head would burst, and he became almost mad. However, he hastened to the well, and bringing up a bucket of water, threw some into his ear. As soon as the cockroach found itself in danger of being drowned, it endeavored to save

Figure 18. Cockroach infestation. Photo © Tony Langford.

itself, and pushed backwards out of the ear, with its hind feet, and thus happily delivered the poor man from his fears."

Roth and Willis also list a number of diseases that have, at one time or another, been attributed mistakenly to cockroaches. Much of the research establishing the presence of bacteria, helminth, and protozoa in the cockroach and its feces was done early in the twentieth century, and many leapt to the conclusion that cockroaches were the primary vector for diseases caused by these agents. In the early 1900s, they were blamed for beriberi, actually caused by a vitamin deficiency. Scientists believed it was caused by either a parasite or an amoeba passed from cockroaches to people. In 1924, at a dinner given by the English College of Pestology, a researcher from Denmark named Fibiger presented a paper postulating that roaches carried cancer and transmitted it to man, and in 1933, a French researcher named Cordier suggested that an amoeba found in cockroach intestines might be responsible for cancer. Others illnesses attributed to roaches, at one time or another, include Bright's disease, malaria, and scurvy.

Despite these false alarms, there is no doubt that cockroaches are capable of passing along diseases to human beings. Even so, most people who have intermittent contact with cockroaches over the course of their lives do not turn up with hookworms or hepatitis or salmonella. It seems, fortunately, that people and *Blattarians* are generally able to live in the same places without transmitting diseases between them. Roth and Willis concluded that the degree of danger to human health represented by cockroaches probably approximates that represented by the common housefly, another insect known to be a vector for disease, with which we have learned to live. They sum it up well: "Cockroaches and house flies are potential health hazards to man because they feed on both human feces and human food . . . Unfortunately it has always been easier to tolerate the cockroach, which shuns daylight, than to ignore the ubiquitous house fly, which breaks bread with us each meal" (Roth, 1957).

Roaches, like flies, can be controlled, but probably never wholly eliminated from our immediate environment. So, if you happen to see a roach scuttling behind a cabinet in your favorite restaurant, should you strike the place from the list of eateries you patronize and never darken the door again? It's a question each person must decide individually.

. . • •

In New York recently, while researching this book, I lodged on the Upper West Side, near 106th and Broadway. Friends in the neighborhood told me that the best cheap breakfast around was at Victor's, a *comida criolla* place right on that corner. It was good, particularly the strong Cuban coffee, which came with eggs, toast, and home fries for three dollars. Equally pleasant was the vivacious waitress whose rapid-fire conversations in Cuban Spanish I enjoyed trying to follow. One morning, sitting at the counter, sopping up the egg yolk with the last of my toast and simultaneously reading the NEW YORK TIMES, I saw a German cockroach out of my peripheral vision, moving along the counter toward my paper. I tensed my arm, preparing to bring the *TIMES* to bear on the insect. The waitress saw it at the same time I did, and in one practiced motion she jerked her order pad out of her apron pocket and brought it down on the roach, sweeping its dying body off the counter with a flick of the pad.

Victor's faced Broadway and shared a back wall with the rear of an apartment building that faced West End Avenue. The maintenance man for the apartments was named Robert, but everyone called him Gigi. Long black hair tied back in a pony-tail, short and round, he often stood outside the building with a beer in a small brown paper sack in one hand, chatting with passers-by. He had been the super for twenty-one years, he informed me when I stopped one afternoon to talk. I asked him

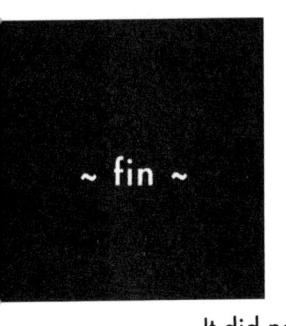

if he had much trouble keeping the apartments free of roaches. It turned out to be a question much on Gigi's mind. Lately, he told me, the building had developed a serious roach prob-lem, and the exterminator had been twice. He said he couldn't understand why sometimes the roaches were so bad in his building, when other apartment buildings in the area didn't have a problem.

It did not take more than a little time sitting at the counter in Victor's, looking down low where the wall met the floor, to fig-ure out why Gigi's building was infested. Sure enough, a look at New York City Health Department inspection reports confirmed my suspicions. Victor's had been closed in November 1997 when an inspector saw "two live cockroaches and several dead ones throughout the establishment," the last straw after years of warnings. Within a few weeks, however, the restaurant had cleaned up and reopened. I'll admit that I did not repeat my breakfasts there, but I did keep going in for the occasional cup of *café con leche*.

. . . ● ●

Cockroaches cause disease, but they have also frequently been used to cure it. The first person, apparently, to address the medical uses of cockroaches was Pliny the Elder, who wrote in the first century that they could be crushed in a poultice and used to cure boils. This was a remedy that was still being listed in Merck's Index early in this century. Bright's disease, dropsy, and diaphoresis are other diseases for which the index listed cockroaches as a possible medicine. In fact, powdered cockroaches appear to have a significant diuretic effect, i.e., they increase urine output. This is, according to Roth and Willis, the one medical attribute that has been medically tested and consistently confirmed.

Pliny also recommends crushed "Blatta" as a cure for itching, scabbing, and for incurable ulcers. These are all pretty far-fetched, but he also recommended grinding the fat of certain "Blatta" with oil of roses and using it to treat an earache. Variations of this cure have been used ever since. There are records of a sixteenth-century Viennese physician using Pliny's cure, and Newbell Niles Puckett reported in his 1926 compendium FOLK BELIEFS OF THE SOUTHERN NEGRO: "An abscessed ear may be cured for life by taking off the head of the cockroach, splitting it in half and pressing the juice into the ear, after which the liquid is held in place with a little cotton."

Historically, people who had less money to spend on doctors and drugs utilized alternative remedies, made from herbs and insects, including, on occasion, roaches. Lafcadio Hearn reported in the NEW YORK TRIBUNE, in 1886, that black people in Louisiana used cockroaches fried in oil for indigestion, as well as cockroach tea for tetanus accompanied by a poultice of boiled cockroaches placed on the wound. In seventeenth-century Jamaica, people drank the ashes of cockroaches as a tonic. They also "bruised" roaches, mixed them with sugar, and applied the concoction to ulcers and cancers so they would suppurate.

In Finland, earlier in this century, it was reported that the cockroach was honored as a protector of life and a bearer of good luck, so much so that they were encouraged to settle and multiply in homes, even those belonging to the well-to-do. In 1886, Miall and Denny reported that ". . . both cockroach-tea and cockroach pills are known in the medical practice of Philadelphia." And, in 1929, the great Spanish poet Federico Garciá Lorca wrote a poem that he turned into a play, EL MAL-EFICIO DE LA MARIPOSA (THE BUTTERFLY'S EVIL SPELL), about a cockroach that saves the life of a butterfly, then falls in love with her, only to be abandoned.

Cockroaches figure prominently in a series of tales heard throughout the Caribbean and Central America. MARTINA Y PÉREZ is a traditional folk tale from Puerto Rico in which a roach named Martina, usually described as a "beautiful Spanish cockroach with black eyes, smooth complexion, well-educated and very proud of her heritage," decides to take a husband. Among the suitors are a cat, a rooster, and a frog, but she rejects them all. Finally, she accepts the hand of the gallant Pérez, a mouse who is so well mannered and handsome that it is thought he is descended from royal blood. They are married, but tragedy soon strikes. While Martina is preparing a Christmas dish in a big pot she steps outside and leaves her new husband in the kitchen. Pérez leans over the pot to sneak a taste and in he falls, leaving Martina a widow.

Anthropologist Jay Melching, in an elegantly written article called, "From archy to Archy, Why Cockroaches Are Good to Think," points out that it is not surprising that some sympathy for the cockroach might be found among people outside of the power structure in the Americas. Domestic cockroaches are believed to have reached the shores of the Americas on slave ships from Africa. Both cockroaches and slaves were forced to make a niche for themselves on these shores. Melching remarks that the cockroach in many Caribbean folktales is assigned a role not unlike that of the Trickster in folklore, one who must survive by his wits in a hostile world. He quotes anthropologist Roger Abrahams who writes of the cockroach in his introduction to AFRO-AMERICAN FOLKTALES: STORIES FROM BLACK TRADITIONS IN THE NEW WORLD: "Like Trickster, he lives at the margins between the family and the wilds, and can be seen as something of a contaminating anomaly, and thus, like Trickster, the upsetter of order."

This ability to survive in precarious, undefined territory is admirable from an underdog's point of view. Melching reviews a number of instances in which African American folklore seems less reflexively harsh toward cockroaches than that of European Americans.

> Perhaps we are prepared now to return to the anthropologist's question, "why are cockroaches good to think?" There may be several answers. Certainly, for most Americans, thinking about cockroaches is a form of reflecting on dirt, and... reflecting upon dirt helps us reflect upon order and disorder, both social and bodily. Thinking about cockroaches helps valorize the positive terms (clean, light, center) that stand in contrast with the cockroach. For this audience of Americans, the cockroach is no mediator of eternally ambiguous cultural dualities. The cockroach unambiguously represents evil. Exterminating the cockroach even temporarily brings an illusion of order to the domestic scene, even if the uneasy realization is that the cockroach will return, perhaps more resilient than ever.

> In contrast, some Americans, usually those at the margins, find the cockroach "good to think" for dramatically different reasons. African Americans and others find the cockroach an attractive icon precisely because he inverts

the normal order of American culture. Moreover, there are important lessons to be learned by reflecting upon the cockroach, important lessons about survival. And, as the Don Marquises and Berke Breatheds and Donald Haringtons suggest, perhaps the additional lesson to be learned from cockroaches, from archy to Archy, is that we should respect the lives and perspectives of even the lowliest creatures, who may have something to teach us.

. . • •

A large percentage of the thinking that most people 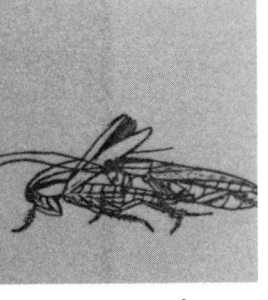 do about cockroaches centers on how to eliminate them, so it is not surprising that a good part of traditional folklore, dating back thousands of years, consists of remedies for ridding one's dwelling of *Blattarians*. An Egyptian papyrus includes in hieroglyphs a prayer to the ram-headed god Khnum for protection from cockroaches. Well over a thousand years later, the Greek scholar Diophanes recommended the guts of a ram, freshly killed and full of dung, buried slightly underground to attract "the Blats." After a pair of days it will be full of them, said the Greek, and you can bury the entrails deep enough to suffocate the roaches, or carry the foul package elsewhere to dump them.

In the modern era, and before chemical companies began marketing their wide range of toxins with which to attack roaches, the preferred method for ridding oneself of the pests seemed to be passing them on to neighbors. In the nineteenth century, the prevailing idea was that if you could slip a person a couple of roaches from your home colony, the rest would follow. From Maryland to Mexico there was a widely recorded strategy whereby a person took an envelope and sealed a couple of roaches and three coins inside. This was left outside, beside the road, and whoever picked it up would be the new owner of your roaches.

The cockroach figures only scantily in traditional, rural American folklore—in all of Newbell Niles Puckett's vast collections of rural folktales, for instance, there are only a handful of roach anecdotes. The cockroach does make an appearance in some "urban myths." One classic is the tale of the woman who left her hair pinned up in a bouffant for such a long time it became infested with cockroaches, causing her tremendous headaches. Among university students, Melching documents tales about killer roaches escaping from laboratories and about roach infestation in the university's student cafeteria food.

Popular culture features numerous representations of roaches, most of them in the time-honored role of the villain, the primeval carrier of filth and darkness, but not all. There is a Cockroach Hall of Fame in Plano, Texas. Roaches figure in the blues (such as, "I'm watchin' a cockroach crawling in an old tin can/when your baby's left you it's so hard to be a man"). There's also the famous song from south of the border, originating with Pancho Villa's soldiers, La Cucaracha, in which the roach is unable to walk because it is missing its two back legs, or because it has no marijuana to smoke, or for one of numerous other reasons. The song has been widely adapted with many versions of its lyrics. And, extending the metaphorical link with the evil weed is the widespread use of the word roach for the short, brown butt of a marijuana cigarette, which resembles a cockroach. Reefer has long been appreciated in urban apartments where cockroaches are not an unknown sight.

There are a number of films in which cockroaches are a central theme, in fact there are enough of them so that they practically constitute a Hollywood subgenre of their own. For instance, the superb independent film director John Sayles raises money to make his own films by writing and rewriting screenplays for more mainstream pics. He did a rewrite of the filmscript for *Mimic,* directed by Guillermo del Toro and released in 1997, which tells the story of immense, mutant insects amassing in the subway tunnels of Manhattan preparatory to invading the city. In an interview, Sayles explained why he declined to be listed in the credits as a screenwriter: "It really was a good basic story. . . . But, finally, it was still a movie about giant mutant cockroaches and people in situations of peril."

Among other recent films featuring *Blattarians* are the memorable exploding cockroaches from the 1997 action film *Men in Black,* the eminently forgettable *Vampire's Kiss* in which Nicholas Cage eats a cockroach, and *Joe's Apartment,* based on an MTV sketch, which chronicles the travails of a young New Yorker who moves into an infested apartment. An earlier contribution to the genre was *Creepshow.* Roaches make cameo appearances in numerous films and television shows, always as visitors from the dark side. Out of unprecedented concern for the rights of roaches, during the filming of *Men in Black,* a representative of the American Humane Society was on the set to count out two hundred roaches when filming began each day, and recount them at day's end. When roaches were crushed on-screen, plastic models filled with mustard were used so that no roach life would be lost.

.   •   •   •

My biggest problem with keeping the hissers in my study was our housekeeper, a middle-aged woman who has a long nose that she likes to keep continually in other peoples' business. She comes once a week to clean our flat in the center of the city, and comes back to this building two other days during each week to clean the flat upstairs and the one across the hall. The first time she came after I returned with Lou Roth's gift, which I had installed in my study, I told her I had brought some beetles back from the United States. She went straight off for a look at them, and her first words were, "They look like giant cockroaches to me."

One November morning, in a disingenuous voice, she said to me, "The woman upstairs told me she saw a huge roach in her kitchen, long and light brown, nothing she's ever seen before, extremely large." She paused a moment waiting to see if I would say something, then decided to go all out. "Maybe one of yours has escaped?"

I waxed indignant. Of course it wasn't one of my roaches. They're all there, and besides I never take the top off the terrarium for any reason, except to change banana skins. She seemed to wonder how I knew they were all there if I never took the top off to count them, but she was too polite to come right out and ask. Just what I need—to be known throughout the building as the guy who brought a strange breed of giant roach from America to infest my neighbors' flats.

CHAPTER 6

war

CHRISTMAS DAY, 1972, found me flat on my back in a tiny hotel room, in a small, southern Moroccan town near the Sahara Desert. I had a raging case of dysentery and was too feverish and weak to sit up for any length of time, so I spent hours just breathing and looking at the ceiling. The hotel, one of only two in town, occupied the second floor over the Café Alaq. Goulimine was a dirt-street town of around fifteen thousand people within fifty miles of the edge of the Spanish Sahara. The iron cots in the hotel looked and felt like they'd been discarded by one of the Spanish army outposts to the south in the desert. Still, conditions were better than at the other hotel, which had no beds at all in its cubicles, only thin mattresses on the floor.

This was one case of dysentery that could not be blamed on cockroaches. I knew all too well where I had contracted it, although it remained incredible to me that something so insignificant as a few drops of water could leave a person, twelve hours later, in such pitiful shape. At frequent intervals my insides forced me up off the cot once again to wend my way past Ali—the hotel's genial, balding, rotund owner who was profoundly at ease in his deep chair behind the rudimentary reception desk—

and down the stairs, across the courtyard, and into the dank, shadowy cement closet behind the cafe that served both the hotel and the restaurant as a bathroom. The toilet, itself, was no more than a hole in a concrete floor over which one squatted. There was, of course, no toilet paper, and to use any kind of paper was to back up the hole and create a disgusting soup, which I would have been the first person to contend with, given my frequent visitation rate. Hygiene was provided with water from a faucet in the wall where people washed their hands after cleaning themselves. Cockroaches crawled across the humid concrete walls. I squatted over the hole, skinny haunches quivering like the proverbial dog trying to pass peach pits, often retching violently at the same time as I relieved myself of yet another thin stream of intestinal fluids.

Goulimine calls itself, "the gateway to the Sahara," and in 1972 it lived up to the name. The town served as one of the early outposts of civilization for trans-Saharan travelers coming into Morocco from Algeria, Mauritania, or even as far away as Mali. The Café Alaq was along one edge of the souk, the town's central marketplace, which every Saturday hosted a large camel market. It rained every day in December for a couple of hours and the souk was always at least slightly muddy. Nomadic Berbers and Tauregs who lived in the desert would come to town to trade in camels, sheep, and silver. The Tauregs were known as "blue men," because the brilliant cerulean dye they used to color their long, hooded robes (called *djellabahs*), eventually stained their hands and faces, as it had been doing for centuries.

I was there to buy glass beads to use in making leather bracelets and necklaces for sale to hippies back in the first world. A handful of people in Goulimine, members of trading families who had long ago stopped wandering and settled in town to do business there, had gained a worldwide reputation as an inexpensive source of these glass beads, which were, in fact, often called Goulimine beads. They were particularly appreciated by European and American hippies for their bright colors and mandalalike designs. The hippies called them love beads. Various other names have been applied to them over the centuries, including trade beads and money beads. They are known to bead collectors by the Italian name of *millefiori* (thousand flowers) and were made on the island of Murano, near Venice. European explorers and colonizers used them like traveler's checks: they were good for commerce just about anywhere, welcomed in trade by the indigenous populations of both Africa and the New World. Over two million pounds of them a year were produced during the last half of the eighteenth century.

In one of those odd about-faces that can happen over time, the beads that had been so avidly consumed by Africans were now being sold back to Europeans, through the wholesalers in Goulimine. These dealers bought them at markets in northern Mauritania and western Algeria where traders had traveled up from further south with them, from Senegal and Ghana and Sierra Leone. One of the largest sources for beads was the vast market at Tinduf in western Algeria. Here, traders came up from western Africa to buy and sell. Goulimine's bead traders met them there and bought their beads in quantity, stuffed into big jute sacks that could have held fifty pounds of potatoes, which they brought back to Morocco packed into the trunks of old cars, or on the beds of trucks. Then they waited for Europeans, or North Americans, or Australians to come buy them.

I had arrived in Goulimine on the morning bus from Marrakesh, the day before I got sick. As soon as I had found the hotel over the Café Alaq and set down my bag, I went down to the souk and began looking for beads. I made myself known, in rudimentary French, to the two or three traders who had beads for sale. Then, I found a small cafe with a back room, where I took my long wooden pipe with its tiny clay bowl, which I had bought in Marrakesh from the same man who sold me the *kif* to put in it, and sat in that room with my back against the wall smoking and sipping fresh mint tea with a half dozen men dressed in *djellabahs* who were doing the same thing. It was my first trip to Morocco, but I had taken to this relaxing local custom right away. My fellow smokers left me alone, and I did the same for them, each of us inside our own mansion of many rooms.

After a while, I went back to the hotel. Word traveled fast in Goulimine. Ali signaled me over to the reception desk as I passed, telling me in fluid French that his family dealt in beads, and that since I was a guest at the hotel, they would make me a good price of about $10 per pound. It *was* a good price considering that with a small length of leather to turn a bead into a bracelet or a necklace, I could get a dollar apiece for them, and

that there were more than one hundred beads in a pound. A tenfold return. For every eight pounds scooped out of sacks, Ali added, I could handpick, bead by bead, two more pounds. Okay, I said, all right, I'll take twenty pounds. He told me to meet him at 4 P.M., and we'd go to his family's house, where I'd still have a couple hours of daylight.

I did. The house was about three blocks away from the souk. The beads were in a half dozen big jute sacks resting on a carpet. Thick cushions were scattered about. A handful of men was sitting around, and Ali introduced me to his brother and his cousins. He went back to the hotel and I joined his family on the cushions. Mint tea was brought in on a tray by a stout, middle-aged, unveiled woman who put it down and left. A hanging scale was produced and the beads were scooped out of the sacks into a brass pan on one side, while weights went into the brass pan on the other. Each two pounds went into a brown paper sack. I took my time handpicking my four pounds from those jute sacks, and no one rushed me. It wasn't easy to choose. Some of the beads seemed more translucent than others, some had remarkable depth to them. Designs, sizes, and colors varied. A few seemed—for some not quite definable reason based on color and feel—to be considerably older than the others.

Once the buying and selling was done, the stout woman reappeared carrying a tray piled with covered bowls. She served us each a lamb *tajine,* that tasty Moroccan dish of meat, dried fruits, and vegetables cooked in a round, shallow, earthenware vessel with a conical lid. With a chunk of Moroccan bread, the *tajine* was delicious. I was happy to be there. Afterward, a bowl of cold water was passed around. How could a person turn down a drink of water at the end of such a meal and not offend the hosts? I raised the bowl to my lips and tried to appear as if I were taking a deep draught when all I did was sip. No more than a few drops slid down my throat, as I restrained myself with a mental image of the bottle of mineral water on the rickety nightstand by the iron cot in my hotel room.

It was barely light the next morning when my stomach woke me and sent me stumbling down the stairs to the Alaq's cement water closet, my insides all knotted up. I passed a pair of miserable days in that hotel before I could force myself out of bed onto a bus back to Marrakesh, and quite a few days more before I was back in western civilization, with my beads, feeling like my old self.

By Christmas of 1997, twenty-five years after my first visit, Goulimine still called itself the gateway to the Sahara, but it was a phrase designed to catch the ears of tourists, just as the camel market only remained to catch their eyes. The same half dozen camels were trotted out each Saturday. There were still a few beads for sale, but they sold by the gram and not the kilogram, and the $20 that used to buy a kilo was enough to buy about a hundredth of that, ten grams, which translates into three good-sized beads. At 1997 prices, a kilo fetched about $2,000. All told, the bead market in Goulimine had virtually disappeared.

The *millefiori* bead trade was a minor casualty of the war in the western Sahara, from 1975 to 1992, between

Morocco and the people of the western Sahara, the Polisario, as they call themselves. There were major casualties: 15,000 men, women, and children died during the fighting, and another 160,000 people went into exile, most of them living in the camps at Tinduf, the same stretch of inhospitable land in western Algeria where so many tons of beads used to be sold at market. When Spain pulled out of what had been the Spanish Sahara, in 1975, Morocco cast a greedy eye on the mineral deposits there and moved to annex it. The local population resisted and a war began, one of those low-intensity wars that goes on and on, away from the spotlight of media attention.

Goulimine was not a place where a foreigner wanted to spend any time between 1975 and 1992. No town south of  the Atlas Mountains was. Morocco had a sizable military budget, and it went into a full-scale war paranoia over its southern border, sealing it off with a thousand-mile-long, six-foot-high mud wall, the approaches to which were heavily mined. An airport was built outside Goulimine, and the village became a military outpost town. While the war was being fought in the desert, the United Nations tried and failed to work out what seemed an endless series of agreements. After almost sixteen years of fighting, a cease-fire was finally agreed on in June 1991, with a referendum scheduled for 1992. The voting was postponed time and again, and by 1996 it appeared that both sides would soon be back at war. James Baker, former U.S. secretary of state under George Bush, agreed to try and bring the two sides together, and in the face of almost universal skepticism about his chances, he brought representatives from Morocco and the Polisario to Houston, Texas, and got a signed agreement with a referendum in December 1998. It did not happen, due to delays in the voter census. Now scheduled for July 2000, it is uncertain whether it will ever happen.

A generation of Polisario were born at Tinduf while the war dragged on, and where beads used to be piled up on the ground for sale in the old market, there are now mile after mile of white and turquoise tents where exiles live and die.

. . • •

Officials in New York City have finally recognized the severity of the city's asthma problem, thanks to Dr. Rosenstreich's study and numerous others, and declared an official "War on Asthma." By 1999, the subways were studded with ads in English and Spanish urging parents to seek treatment for their child's asthma, a citywide hot line had been established, all elementary and middle school teachers received asthma training, and a campaign was designed to teach children not to be ashamed of the disease, but rather to deal with it, to use their inhalers and avoid behavior that might cause an attack.

Nevertheless, at just about any moment of the day, in any of the city's hospital emergency rooms, there is a child sitting hunched over in the grip of an asthma attack, waiting to see a doctor, all energy focused on the struggle to breathe. Many of these children make multiple visits each year to their local emergency rooms, and while the situation may become familiar to a child, an asthma attack is always terrifying.

Despite the highly trumpeted new commitment from the New York City Department of Health to combat asthma among the city's poorer children, there is little or nothing being done to remove cockroaches from public housing. Without a lot of time and dedication, it is nearly impossible to keep an apartment in New York City public housing free of *Blattella germanica*, the ubiquitous German cockroach. And, even those who keep a clean household have to maintain constant vigilance against spillover populations from their less scrupulous neighbors. There are more than forty exterminators working full-time for the New York Housing Authority (NYHA), and the city spends nearly half a million dollars a year on insecticides, but the roach problem in public housing is not improving. In fact, it's getting worse, according to residents and exterminators.

As it happens, this is just the opposite of what is going on in many private New York apartments, where roach infestations are on the wane, after centuries during which the city's *Blattarian* residents multiplied and flourished. For almost as long as Europeans have lived in New York, the city has had roaches. As early as 1747, Swedish botanist Peter Kalm wrote: "They [cockroaches] are in almost every house in the city of New York." By the early 1800s, a popular novelty item in the stores of New York was a roach-shaped sugar candy. As the city became crowded with immigrants, tenements swarmed with roaches, and as the human population increased, the *Blattarian* did likewise.

Despite such a long and persistent history, the city's present roach population is declining. While no roach census has been tallied—an impossible venture—both residents and professional exterminators agree that things are improving. Many long-time city residents will tell you that over the past couple of years they have not seen the numbers of roaches that have been, for centuries, accepted as a necessary price to pay for living in many neighborhoods throughout the five boroughs. "Not too long ago, I had so many cockroaches that at night the walls moved where they had gotten in behind the wallpaper and were eating the paste," a friend, who lives on the east side of Manhattan's Midtown, told me. "I haven't seen a roach for a couple of years now."

The awful moment of reckoning that most New Yorkers have faced—turning on a light to the mad dash of scores of little brown bodies scattering for cover—may be, for many New Yorkers, a thing of the past. The reason, although most New York residents may not generally be aware of it, is the widespread use by commercial exterminators of a new form of pesticide. It is a gel bait, virtually nontoxic to humans but fatal to roaches. It has become the weapon of choice against cockroaches, largely displacing the old organophosphate sprays and pyrethroid-based powders, and, by most reports, doing a better, safer job.

That should be good news for the asthmatic children in public housing, and it would be except that unlike numerous other housing authorities around the country, the NYHA has so far refused to switch to the gel. The department's exterminators continue to use sprays, even though the state's Environmental Protection Agency required the NYHA to stop routinely using organophosphate sprays in 1996, which were, until then, the roach killer of choice. Currently, a wettable pyrethroid powder, called FICAM, is mixed with water and sprayed. While less toxic than the organophosphates, it's not something you would want to ingest, and residents are still required to cover their dishes and cooking utensils when their apartments are treated.

"We are employing a lot of different methods," said Jamal Rasheed, the NYHA's technical advisor for pest control in its approximately 150,000 public housing units. "We don't have a contract yet for the gels. I don't think they're really better, and I certainly wouldn't recommend eliminating sprays."

However, Dan Dickerson, director of pest control for the city's Board of Education—responsible for keeping roaches out of classrooms in approximately 1,200 schools—has done just that. "We haven't sprayed classrooms since 1993. Up until then it was spray, spray, spray.

"What happens is that if you go into a building with multiple rooms, or apartments, and spray, you may kill 90 percent of the roaches, and the survivors scatter. Even the pyrethroids are somewhat repellent, so the roaches spread out and infest areas that were not infested.

"With the gels, it's the opposite effect. They are not repelled, nor driven to move. We leave enough bait to kill them all. I've been in this business thirty years, and the gels have made a tremendous difference. You do the job once and it's done. Complaints are down 80 to 90 percent. It used to be roach, roach, roach, but now mice are a bigger problem."

.   .   •   •

Nashville, Tennessee's housing authority is one of the numerous public housing authorities across the country, unlike New York's, that has made the switch to gel baits. Its exterminators use a product called Siege, manufactured by American Cyanamid, which uses a chemical called hydramethylnon as its active ingredient. A thin bead of what looks like soft wax is laid down from a tube, using a device much like a caulking gun. This is done in cupboards, along baseboards, near drains, and wherever roaches like to gather. The gels have an extremely low toxicity, and dishes can be safely stored in the same cupboards where it is used. In addition, after a roach ingests a little of the gel bait, its feces become toxic to other roaches that may practice coprophagy, or the eating of feces, a standard practice among German cockroaches (Silverman, 1991).

I went along one morning with Booker Washington, fifty-three, a supervisor of pest control in Nashville's housing projects for over a dozen years. He's a solidly built, African American with salt and pepper hair that's cut short. He always calls cockroaches bugs. "Before we stopped spraying and started using the bait, a year and a half ago, the bugs were pretty much out of control," he told me. "Now, they're way down and people are much happier because they don't have to get all their dishes out of the cupboards and cover them up before we come."

Nevertheless, before we entered the first of the forty housing units he would visit that day in the James Cayce Homes, as he stood knocking on the aluminum frame of a screen door with only one tatter of torn screen left in it, he warned me, "This one's bad, but don't worry. They're not all like this."

A young, obese, white woman with stringy brown hair and puffy eyes answered the door. In a low voice, she told Washington that her child was sick, and that he should come back some other day. He explained that she had received a notice two days before that he would be coming, and if he couldn't come in he'd have to notify the housing project's supervisor. Reluctantly, she stood aside to let us pass, muttering, "I hope you don't get whatever it is I've caught from my little girl. We're both real sick."

My stomach turned over walking into the twilight inside. The smell was gag strength, compounded of cat shit and garbage, with an underpinning of the bitter, malodorous aroma that denotes an infestation of German cockroaches. The floor was covered with loose pages of old newspapers; cat food and kitty litter crunched underfoot; and in the kitchen was a fifty-five-gallon plastic garbage can, overflowing with paper, half eaten pizzas, empty Coke cans, and a whole lot more.

Roaches were everywhere, on every countertop, scurrying across the walls and the floor. Booker Washington got right to work, pulling the gel applicator gun from a holster on his hip and opening the cupboard doors under the sink. On the wall of the kitchen was a photocopy of the Serenity prayer:

> God, grant me the serenity to accept the things
> I cannot change,
> The courage to change the things I can
> And the wisdom to know the difference.

Both of us were in a hurry to get back outside, and within five minutes the kitchen and bathroom had been treated. "Thank you ma'am," said Washington, as we moved briskly out the front door. She said nothing, staring at him with a dull resentment. We stood outside on the front stoop a moment, breathing deeply. Washington was wearing thick-soled, black work boots, even though it was hot summer in Nashville. He suddenly broke into what looked vaguely like a buck-and-wing, pounding his boots on the stoop.

"They call this the bug people's stomp," he laughed. "It may look funny, but they *will* get on you. You'd be surprised how easily you can carry some home from a place like that."

In the next unit we entered, an overweight black woman, with no front teeth, wearing a kind of torn nightshirt, sat on the couch watching television as Washington worked in her kitchen. She had two ragged-looking Pekinese dogs on the couch beside her. On the TV screen were a pair of elegant white women, one showing the other how to make a Sazerac cocktail at a gleaming bar in a television studio, explaining that it was a drink concocted and made famous in New Orleans. This was immediately followed by a commercial break advertising Martha Stewart's magazine, *LIVING*, a "compendium of good taste and good manners," according to the announcer. Above the TV, childish drawings were Scotch taped to the wall, and signed in big sloping letters: Deshen, Jamika, Jermaine, Latanya.

Unbidden, the resident began to talk about the gel bait and what an improvement it was. "He used to use a spray," she told me, nodding toward the kitchen where Washington was opening and closing cabinet doors. "The roaches would fall out and die all over, and then they'd come right back. But this stuff they're using now is great.

"Honey, it used to be bad, really bad. My niece came over to visit me once and when she got ready to go, she picked up her purse and the roaches ran all out of it. I mean, it was really embarrassing. Now, with that bait, you don't ever hardly see a roach."

The most important thing about the gel bait, said Washington, is that it works, and that a resident who keeps a relatively clean housing unit can be virtually roach free for the three-month interval between treatments. A secondary benefit, he said, is the change in attitude it has engendered in many residents.

"Before, when they saw you coming with that spray, folks just weren't very happy to see you, particularly old people with breathing trouble or mothers with young kids. They didn't trust that stuff at all, and it didn't work well, either. Now, most folks are really glad when we come by and they appreciate what we're doing for them. That makes a difference."

. . • •

It's a difference that Eddie Slade, in New York, would like to experience. Only his Christian faith and prayer get him by in the face of some of the nasty treatment he receives at the hands of the tenants of Bronx public housing. He has been an exterminator with the NYHA for fifteen years and has spent the past two of them working exclusively in the Bronx.

"I'm a born-again Christian," Slade told me in his tiny office in the basement of the Gun Hill housing project. He is forty-five, a short and solid black man, born in Brooklyn, who wears a mustache and a trimmed goatee. "That means I have to go to God and ask for forgiveness every day, and so I also try to give it to the people I deal with. Yesterday, for instance, I had to go to 10-D, and I know how mean they can be to me sometimes, so I asked God to help me before I knocked. I said, 'God, you know how this tenant is, please help me be patient.' Still, it can really get to you."

It is in this little basement office that he begins each day's work, which consists of some thirty-six apartments to be treated every day, at housing projects all over the Bronx. He starts each morning with paperwork and a prayer, then mixes the first one-gallon stainless steel tank of FICAM powder with water, picks up the tank, and sets out on his rounds.

Unlike Booker Washington, he was waging an uphill battle, facing increasing numbers of cockroaches each year. Left to her own devices in a Bronx apartment, a female *Blattella germanica* may live close to a year and produce about 2,500 new cockroaches during that time. Eddie Slade said their numbers were growing yearly. "We have more and more of them. One reason is that they used to incinerate the garbage, but the environmentalists decided that was bad so now it sits out until a Sanitation Department truck picks it up. That's a change that really favors the cockroach," he said, as we stood outside a seventh-floor apartment waiting for a response to his knock.

A tall, disheveled woman opened the door, a baby in her arms, a toddler down near the floor hanging on to the bathrobe she had around her. Both kids had snotty noses; nobody looked healthy. The shades were down throughout the apartment, the linoleum on the floor was stained and torn. I could smell the roaches in the kitchen even before Eddie hit the light switch and they scattered off the dishes piled in the sink. The tenant had been notified two days before of Slade's pending arrival, but she had not removed her dishes, pots, pans, and utensils from the kitchen shelves and covered them up as the notice requested she do. Eddie Slade sprayed anyhow, taking care to aim away from them.

"Miss?" he said to the woman who stood at the kitchen door, balancing the baby on one hip and watching us, an annoyed look on her face. "Miss, you've got a real roach problem here."

She barely nodded to indicate she had heard him speak the obvious. "I'm a pretty good exterminator," he told me as the door closed behind us, "but I'm not a miracle worker. Jesus was a miracle worker, but all I can do is do my job. I mean I can only go with what I have, and a person's cleanliness is still the most important thing. I keep up my end of it. I take classes—not long ago I took a whole day's worth from this one guy. Man, that guy loved this work. He went on and on about exterminating. I forget his name but he's one of the best."

. . . •

I know his name. He is entomologist Austin Frishman, one of the nation's most highly respected pest control experts, who taught for twenty-three years at the State University of New York at Farmingdale and has his own pest control business, which he runs from his Long Island home. His work constantly takes him around the country to deal with tough pest problems, and he gives regular seminars. He has probably taught, at one time or another, most of New York's exterminators, and his book, THE COCKROACH COMBAT MANUAL, is used by professionals and homeowners alike. He was one of the early developers of cockroach baits, and is a consultant to the Clorox Corporation, maker of the popular Combat bait stations, which are widely used by consumers. There were more than 185 gel baits on the market in 1999, according to PEST CONTROL magazine, all developed within the past decade, most of them for consumers. Professional exterminators tend toward either Siege from Cyanamid, or MaxForce, marketed by Clorox.

"The New York Housing Authority is behind the times and it's an unfortunate situation," Frishman told me, as we drank Dr. Brown's cream soda on a balmy summer evening on the deck beside his backyard swimming pool. "It used to be accepted that the way you controlled cockroaches was by bombs, sprays, smelly stuff where you've got to take everything out of the cabinets, and that's not so anymore. The new bait technology is very effective.

"Getting a governmental body to change is like trying to turn a very large ship. We're trying to get them to use baits, but they're numbers pushers. It's actually cheaper and more effective to do it with baits. It takes longer initially to get a bad situation under control, but it works and you don't need to come back as often.

"It's hard to get them to understand that. Their guy comes in with the spray and waves it around the baseboards, and then he goes out. You have to put the bait in the specific places where they're coming in. It takes a little more time. Not a lot. A little. But the city official says, 'I have a man and my production is forty apartments a day, and I don't want to hear about baits.'"

Extensive test results have been presented to NYHA officials, by other housing authorities and the manufacturers of the two commercial gel baits, but the exterminators who work for NYHA are still carrying spray tanks, instead of gel guns. Roach infestations continue unabated, while the numbers of inner-city children with asthma continue to climb. In the Bronx, some 8.6 percent of the children suffer from asthma, more than double the rate for those in other urban areas. Frishman was, nevertheless, confident that sooner or later the NYHA would see the light and realize that pest control had changed, and that they must change with it. "We know how to get rid of roaches now, we have the material, and we need to use it and develop a policy of abatement."

Cockroaches have been good to Austin Frishman. Not only is he known by most of New York's exterminators, but he was the subject of a documentary by National Geographic called *Doctor Cockroach*. He has appeared on numerous national television shows talking about roaches, and has consulted with the rich and famous about *Blattarians*. He describes himself as semi-retired, but once a month or so, his phone will ring and a desperate restaurateur with a roach infestation that local exterminators can't remedy will offer him a ticket and a lot of money to come and clean out the premises. "I wouldn't want to stop working completely," he told me. "I like what I do. Besides there's no way I can get away from it. I have a secret e-mail, but people find me anyhow. Just last month I went down to Louisville because there was a guy there who just didn't have anywhere else to turn."

Every year, the Clorox company holds a contest, which they announce in publications like THE NATIONAL ENQUIRER, for the house or apartment with the worst cockroach infestation in the United States. The winner gets $5,000 in cash and the free services of Doctor Cockroach. In 1997, more than a thousand people entered the contest. The winner was not from an inner-city, but from suburban Atlanta. Mary Esposito's winning letter read in part: "When I turn on my oven roaches swarm all over—you can see them trying to get out through the clock on top of the stove. If I don't cover the pots when I'm cooking, they fall from the overhead fan onto the food."

Austin Frishman was sent to Atlanta, and, with the aid of seventy-two Combat bait stations, he cleaned up the 75,000 roaches he estimated were living in Mary Esposito's house. In a lifetime spent battling cockroaches, he has witnessed some mega-infestations. "Cockroaches are omnivorous and ubiquitous, so they can get very, very bad in lots of places. In restaurants, I've seen it rain cockroaches for twenty minutes after we've treated. I've seen it like that in apartments. I've seen private homes so bad that the roaches were outside in the shrubbery and trees because there was no room inside—way, way over a million.

"I've seen American cockroaches in hospitals, and there were so many of them they were on each other's backs. I've seen them get into dialysis machines. I've seen people with no eyebrows because they were chewed by cockroaches. In America. I've seen little babies that have been bitten by cockroaches. I get these types of calls on an emergency basis. Companies around the world call me when they have a really bad problem. It may be a ship, a submarine, an airplane. They call, and I'm on my way."

The great majority of pest control operatives don't find themselves jetting around the globe killing roaches. They get up and go at it every day in the communities where they live, take their cholinesterase tests on a regular basis, and deal with all the vagaries and ill humors of citizens who are not happy to be spending good money on getting rid of bad bugs.

The cholinesterase tests are just another reason why the gel baits are catching on among commercial exterminators. The organophosphate sprays, used for many years, kill bugs by inhibiting the production of cholinesterase, an enzyme in the blood that is essential for the proper working of the nervous system. Chronic inhibition of the enzyme causes twitching, convulsions, and, eventually, death in both insects and humans. Exterminators who use organophosphate sprays are advised to have regular blood tests to make sure that their cholinesterase levels are up where they should be. The primary organophosphate in the sprays is a chemical called chlorpyrifos, and as with many toxic chemicals, those most at risk from its side effects are the young and the old.

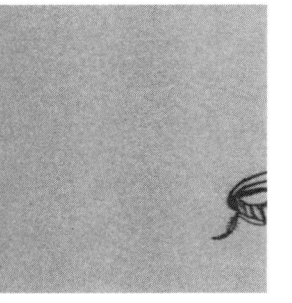

Researchers at the Center for Disease Control took random urine samples from one thousand adults in 1995 and analyzed them for pesticides. Chlorpyrifos was present in 82 percent of the samples. In a report on chlorpyrifos hazards, called UNREASONABLE RISK, authors from the nonprofit Center for Public Integrity wrote: "There's no escaping chlorpyrifos. Americans use 11 million to 17 million pounds of it a year. On any given day, you can find products containing it in 20 million U.S. households. It's the main ingredient in Raid, d-Con, Dexol, Enforcer, Ortho, and more than nine hundred other products. It may even be in the shampoo you use on your dog. In 1993, DowElanco, a joint venture of Dow Chemical, Inc. and Eli Lilly & Company, sold 27 million pounds of chlorpyrifos worldwide, under its commercial name, Dursban.

"Once sprayed, chlorpyrifos can cling to carpets, countertops, furniture, and toys; envelop our children and our pets; lace the food we eat; and pass through our skin. Children are particularly susceptible to it. . . ."

Companies like DowElanco have come under increasing public pressure over the past few years regarding the toxicity of chlorpyrifos. Charles Lewis, from the Center for Public Integrity, summed it up in his June 1998 testimony before Congress on why the pesticide industry needed to be more closely regulated: "For decades during the Cold War we talked about how if things went too far the only species that would survive are cockroaches. Is it far-fetched to imagine that the thing that kills them could also do some harm to humans?"

Oddly enough, such pressure for increased regulation has resulted in chemical companies paying human beings to act as guinea pigs in tests on chlorpyrifos and other cholinesterase inhibitors. In 1996, the Environmental Protection Agency toughened its standards on results obtained from animal testing. Prior to that date, the Food Quality Protection Act mandated that any pesticide causing adverse effects in animals could only be present in food in one-hundredth the dose required to damage an animal. That was raised to one-thousandth under the new regulations. The chemical companies reasoned that the best way to overcome that was simply to skip the animal tests and use human guinea pigs to demonstrate the innocuous nature of their products. The EPA accepts such data, and there has been an increase of testing on humans to provide evidence that a given pesticide is safe.

In a group of tests in England, test subjects were offered money to ingest capsules containing small amounts of pesticides, and the results of these experiments were accepted as valid by the EPA. Among the substances for which tests on humans are planned in the U.S. is chlorpyrifos, according to the WALL STREET JOURNAL.

"These pesticide experiments are being conducted on humans abroad, then accepted by the U.S. government in the absence of specific EPA regulations or monitoring capacity for human research," according to a report from the Environmental Working Group, another nonprofit watchdog organization in Washington. "These companies are not testing medicines on people to see if they are therapeutic. They're testing toxic chemicals to see how high exposure can be without causing regulatory problems. No one ever benefits from being exposed to pesticides."

Despite their dangers, organophosphate-based sprays are still in use, particularly for large jobs or in places where there are major infestations. Even in apartments, they have advantages, mostly economic, for the professional exterminator. In New York City, for instance, landlords are required by law to have their buildings treated for cockroaches every thirty days. Many of them simply ignore the law. Those who choose to comply with it contract an exterminating company. The going rate for such a contract is $40 a month to treat an entire building.

"That's the price for a building with fifty apartments, but it should really be hundreds of dollars, if you're going to do it right," said William Garcia, the energetic manager of Acme Exterminating Corporation, a midsized company in Manhattan. Garcia has been in the business for twenty-one years. "It's too cheap. You can't spend enough time in each apartment. You go during the day and most people are out working. You go in, hit maybe three apartments, maybe hit the basement, and you're out in fifteen minutes. The sprays cost less than the gel, and they're quicker to use, so you still have a lot of companies that have not switched over.

"We have. Gels are the best thing that ever happened to pest control. We've been using them for the past five years. We do Presbyterian Hospital here, which is one of the biggest hospitals in the country. When we took it over there were four technicians up there averaging maybe sixty service calls a month, besides the regular service. When I went in we went to the gel and an integrated pest management program immediately. Service calls went down to five or six a month."

Integrated pest management (IPM) is the new buzz phrase in the pest control industry, and it means doing everything with maximum efficiency and minimum toxicity. It is what

Dan Dickerson is practicing in New York City's public schools. In addition to baits, it involves using silicon sealers to close off the cracks and spaces that allow roaches to move from one harborage to another; making sure that all leftovers from previous infestations, such as feces, carcasses, or molted skins, are vacuumed away; and being careful about leaving any food or crumbs around. It is holistic extermination, paying attention to all the variables, rather than just going in and spraying heavily on a frequent basis. Although many professional exterminators would deny it, IPM has come about, mainly, as a response to increasingly strict environmental regulations imposed on the industry. Garcia applies the same kind of IPM program on a minilevel with his private clients. For about two hours' work, at $75 per hour, he'll rid an apartment of cockroaches and guarantee it will stay that way for a year. The initial consultation is free.

Insect growth regulators (IGR) are another recent addition to the antiroach defense system, particularly those derived from the juvenile hormones that control a cockroach's development toward adulthood and reproductive capacity. When IGRs are sprayed, they will cause German cockroach nymphs to develop into what are called "adultoids." These insects are adults, although they often have some external deformities like twisted wings, and they go through all the motions of courtship, such as antennae fencing and wing raising, but they are unable to successfully copulate and pass sperm. The hormones are called juvenoids, and while they have the advantage of not being toxic to humans, they have the disadvantage of requiring months in order to significantly reduce an infestation and of a short residual life (Kaakeh, 1997). Many pest control operators use IGRs as the last step in their integrated pest management programs to make sure that any roaches left after the treatment will not be able to reproduce and reinfest the site.

The major problem that exterminators have with gels and IGRs is that they have no immediate knockdown effect. John Wickham, an English pest control consultant, defined knockdown as: "The inability of the insect to move in a sufficiently coordinated manner to right itself and progress normally and does not equate to paralysis which is the loss of movement." That is to say, knockdown is a roach on its back, struggling to turn over, its six legs waving helplessly in the air. When a roach eats a gel bait, it heads on home before the active poison reaches its foregut and destroys it. Customers who are paying $75 an hour like to see those roaches struggling to get up, in agony and convulsions, and the sprays, with substantial knockdown effect, provide them that gratifying visual reassurance that their problem is being solved and that they are getting their money's worth.

Still, the sprays are losing favor, because the gels appear to be a safer, more effective, treatment. The baits are 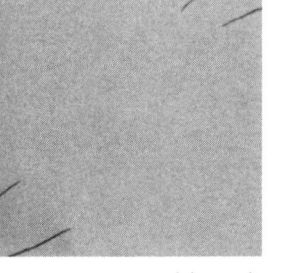 making a dent in New York's cockroach population, which long-suffering residents welcome. But no one is leaping to the conclusion that the use of baits will finally control and eliminate cockroaches. Exterminators have, in fact, had some trouble with the gel along the way. Shortly after gels were introduced on the market, it was found that the sugar being used to formulate the baits was acting as a repellent to cockroaches, a problem that was solved by substituting fructose for glucose (Silverman, 1993). So far, roaches do not seem to have developed any resistance to the gel, or distaste for the fructose, but it is, certainly, only a matter of time.

"In the battle of man versus cockroach, my money is on the cockroach," Austin Frishman told me.

It is no wonder. For years now, random collections of German cockroaches have turned up some that are resistant to almost all the known insecticides, including pyrethrins, organophosphates, and even DDT (Cochran, 1989), which held such great promise in the 1950s as a pesticide, but was eventually banned for the dangers it posed to animals and humans. Experiments have shown that resistance in cockroaches can be inherited (Ebbett, 1997). Once a cockroach develops a resistance, it is encoded in a gene. As generations reproduce, resistance becomes a part of the animal's genetic code. The insect's ability to metabolize the poisons in the insecticide grows, so that increasingly large dosages are needed to reach knockdown and kill levels. This, for instance, has happened in a strain of chlorpyrifos-resistant German roaches. They produce an increased amount of an enzyme that permits them to metabolize more of the poison.

In addition to this biologically inbred avoidance, roaches learn, fairly quickly, to stay away from areas where insecticides have been sprayed. This learned behavior is referred to as the "repellency" effect of insecticides. Given a choice of entering a dark and, normally, attractive area that has been sprayed, or an illuminated area free of pesticides, they quickly begin choosing the latter, despite their natural disinclination to be in well-lit places (Ebeling, 1966). This is called associative learning and was described by entomologist Roland Metzger as "a relatively highly developed type of learning for invertebrates." That cockroaches possess the capacity to use this highly developed learning does not bode well for the possibility that humans might one day be able to eliminate them.

There is no way to know how the first humans to find cockroaches in their dwellings reacted, but it is probable that it was in much the same manner we do. Over the centuries, a wide variety of methods and means have been used in an attempt to kill roaches, from squashing them with a nearby rock to the gels preferred today. These have met with more or less success, but nothing has ever come close to eliminating our fellow travelers, nor is anything likely ever to do so. Over time, roaches have proven quite capable of adapting their bodies and their behaviors to whatever people can throw at them, and when genetic modifications have been required to continue thriving in the company of humans, it has not taken many generations to make the necessary adjustments. Almost as soon as an effective poison—spray, bait, or powder—goes into widespread use, cockroaches begin to develop resistance. And, typically, the products that do the best job, turn out to be more detrimental to our own health than to the roaches.

After so many centuries of trying, in vain, to eliminate cockroaches, there still persists a sense that if among nature's myriad forms there is something as repulsive as a roach, then something must also exist that will kill it. A wide range of products have been proposed, promoted, and sold for that purpose, some of them mildly effective, others completely useless. Boric acid is a cockroach remedy that people have been using since the mid-nineteenth century, and it is moderately effective when dusted around baseboards as a powder. Unlike stronger insecticides, it does not have a repellent effect, and the insects never learn to avoid it. Despite the length of time it has been in use, the exact mechanism by which it kills roaches is not clear. Probably it damages the foregut, and almost certainly works both when roaches ingest it while cleaning themselves, and when they come in external contact with it. Experiments have proven that it is capable of penetrating the cuticle of the insect's exoskeleton. Both American and German cockroaches that have their mouth parts sealed with wax still die when exposed to boric acid dust (Ebeling, 1975).

Another powder with essentially the same action, and which was meant to supplant boric acid in the first half of the twentieth century, is sodium fluoride, which proved to be considerably more toxic to humans. Its use has virtually disappeared, while boric acid continues to be marketed and used. One of the primary disadvantages of boric acid is that it does not have an instant knockdown effect. It takes about ten days to substantially effect a colony and does not have the immediate results that professional exterminators need to satisfy their customers. When applied in the right places, and kept dry, boric acid is an effective roach killer. While it is less toxic than a number of other insecticides, it is toxic, and numerous reports of boric acid ingestion are made to poison centers around the country each year. Infants, in particular, may come in contact with it while exploring their immediate environs, and it can be lethal (Linden, 1986).

There are a range of products marketed to kill *Blattarians* that have no toxic effect at all on humans. Unfortunately, they usually don't work. Among these are electromagnetic devices that purport to alter the geomagnetic field around a roach and thus disrupt its feeding and mating rhythms, as well as electronic vibrators that use tiny electrical vibrations to accomplish the same disruption. Tests run on all these sorts of units have consistently failed to find them in the least bit effective. Another similarly useless method of extermination is ultrasound, which purports to emit tones that are above the range of human hearing, but which drive cockroaches crazy, so they have to leave or die. Such units have been on the market since the early 1970s, and can still be found offered for sale in the pages of reputable magazines. The appeal of such a product is obvious: plug it in and it goes to work without any disagreeable smells, tastes, or noises. Consumers have spent a lot of money on these devices over the past twenty-five years. The only drawback is that in test after test scientists have been unable to demonstrate that ultrasound devices are the least bit effective in controlling roaches (Rust, 1995).

Among the other alternatives to chemicals like pyrethrin and chlorpyrifos are Osage oranges, called the "mock orange" in the south. Supposedly, placing them in strategic corners of the kitchen and bathroom will repel roaches. Powdered pyrethrum flowers are also marketed as an alternative to the chemicals that are, themselves, based on the pyrethrum formula.

There is also the "biological" method of control. In New York City, an example of this is the popular gecko lizard, sold in many pet stores for around $20. The gecko has an insatiable appetite for roaches, and many people have reported that after bringing home a gecko to their roach-filled apartments it was only a matter of months before they had to start buying food for them since all the roaches had been eaten. Geckos, like roaches, are nocturnal. During the day they find a resting place, out of sight, and at night they come out to feed. One of their few inconveniences is that they make a considerable noise as they crunch the roaches between their teeth. People who have a serious infestation of roaches have reported that after a little while they came to take a great deal of pleasure from the sound—presumably finding it soothing in the same way many do the sizzling noise made by a backyard bug light as it fries mosquitoes.

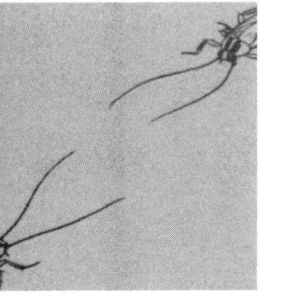

There are other natural enemies of the roach that have been enlisted in the battle to control them. Travelers in the eighteenth century reported that spiders were kept by Jamaicans in their houses because they kept them roach free. In the nineteenth century, hedgehogs were supposedly kept in English homes for the same reason. And, more recently, parasitic wasps have been tried. There are a number of species of these wasps, that lay eggs inside cockroach oothecas. When the wasp larva hatches, its first meal is cockroach eggs, and it will devour all of them before leaving the empty egg case behind. There are situations in which the wasps are the weapons of choice—when, for instance, an infestation is dispersed. In the summer of 1998, the University of Colorado had an infestation of cockroaches in the steam pipes and crawl spaces underneath a number of the buildings on campus. They paid some $11,000 to Barry Pawson of North Ridgeville, Ohio, for 12,000 female wasps, and, according to news reports, pronounced themselves satisfied with the results. Pawson is the only person currently marketing *Aprostocetus hagenowii* in the U.S., and because the wasps are tiny, inoffensive, and short-lived they hold promise for some situations.

Another intriguing possibility for nonchemical cockroach control is the application of stress. In stressful situations cockroaches produce an auto-toxin. If you keep applying stress long enough, the auto-toxin paralyzes the cockroach, and even when the stress is removed, the insect does not recover, and shortly dies. This phenomenon was first noted when insects were exposed in laboratories to DDT. Then, other kinds of stressors were used that were not, in themselves, lethal. For instance, cockroaches were continually "tumbled" for two hours by being put in a jar that was constantly rotated (Cook, 1974). Afterwards, well over half were paralyzed and never recovered the use of their limbs. This same paralysis has been noted in cockroaches that have been immobilized for some hours by being tied down with cotton thread (Heslop, 1959). When they are finally released, they cannot move.

Of course, for your average citizen who has to go to work as well as wage war against the cockroaches in the kitchen, this option does not have a lot of immediate applicability. Most of us are not able to tie down large numbers of cockroaches, nor spend a couple of hours tumbling them. Nevertheless, the fact that there is a toxin within the bug's own body that can kill it provides an interesting direction for further research.

. . . •

One method that works quite well to control German and American cockroaches is to freeze them to death. Humans can survive conditions in a heavy sweater that will put every cockroach in a house into suspended animation, and most pest species simply cannot survive in real cold. For instance, the mortality rate at 15 degrees Farenheit is 100 percent. Cold is also, I'm sorry to report, pretty effective with a colony of Madagascar hissers, and at considerably warmer temperatures, judging by my own experience. Barcelona is full of flats like mine—high ceilinged and draughty, without central heating. People survive by wheeling space heaters from room to room. Powered by stubby orange tanks of butane, they are sold by undocumented foreigners who are the only ones willing to put the thirty-pound tanks on their shoulders and trudge up the stairs in a building without an elevator. My study is unheated at night, and it gets cool. The roaches grew increasingly sluggish and, one by one, turned up dead as the winter wore on.

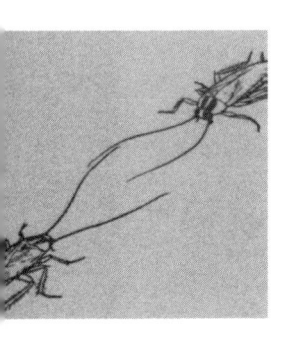

coexistence

COCKROACHES HAVE DIED FOR MY SINS.
There was a period in 1971 when I came back to Nashville, discouraged and defeated by the larger world outside, looking for nothing more than a place to lick my wounds and a routine day job. Both were easy enough to find. I took a job at a bookstore, and rented a two-room basement apartment with walls of stone and a little rectangular window at eye level in each room, through which one could see the feet and legs of people passing by on the sidewalk. It was always deep twilight in that apartment, even during the day, and slightly damp. A perfect *Blattarian* environment, which they had not overlooked. The place was teeming.

There was no television. I furnished the stone rooms with second-hand motel furniture purchased at Goodwill and a record player. Each night after work, and a quick meal at a restaurant, I would come home and sit listening to music, sinking further and further into gloom. I was so unhappy that I could not bring myself to cause suffering in any other life, and I left the roaches to their frolicking. I spent many evenings watching their activity, particularly in the bathroom, where at night, even with all the light cast from a bare bulb, they ran across the walls and the sink. I noted the appearance of nymphs and watched them go through stages of growth, fancying that I recognized individuals night after night. I experimented with music, putting on such diverse sounds as John Coltrane, Bob Dylan, and Ludwig van Beethoven, watching for differences in the roaches' behavior, unsure whether they were actually reacting to the music or whether I was imagining it. The bathroom became a sort of aquarium, and I watched the roaches, mesmerized, as one watches exotic tropical fish, unable to read the order therein, but convinced that one existed.

Finally, some months later, when I could not bear to sell another illustrated Bible, diet book, or paperback mystery, I gave

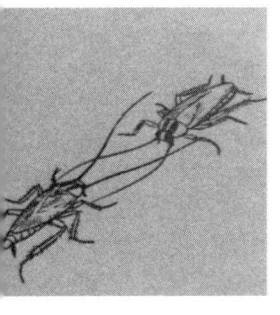

notice at the bookstore, and the same night, on the way home, I bought a can of Raid spray. The next morning, for the first time in many months, I made my bed. Some crisis had been resolved, whether for better or worse, it was hard to tell. It still is. At any rate, I was ready to get on with my life. I was soon gone from Nashville, and it would be years before I returned.

. . • •

Hunting has always seemed to me a cruel pastime and my habit of giving names to animals—something I've done ever since I was a child—prevents me from ever doing any harm to any creature, however repellent. Imagine, for example, that you have a cockroach living in your house and one day it occurs to you to christen that cockroach José María, and then it's José María this and José María that, and very soon the creature becomes a sort of small black person who may turn out to be timid or irritable or even a little conceited. And obviously in that situation you wouldn't dream of putting poison around the house. Well, you might consider it as an option but no more often than you would for any other friend.

from *OBABAKOAK*
by the Basque writer Bernardo Atxaga
translated by Margaret Jull Costa

● ● ● ●

Perhaps so, but it does not often happen like that. If one were to compile a list of the most detested members of the animal kingdom, the cockroach would be prominently positioned near the top. A Yale environmentalist has conducted surveys, among people from all social classes, on this subject. While a number of animals appear close to the top of the lists, including mosquitoes, rats, rattlesnakes, bats, and vultures, cockroaches consistently wind up at the pinnacle (Kellert, 1998). Likewise, in a 1976 survey among Brooklyn fifth graders, roaches came in first as the scariest animal, followed by rats.

There is a psychological syndrome, called delusory parasitosis, which refers to the delusion that there are parasites or insects under the sufferer's skin. The syndrome's definition is often expanded to include the conviction that one is living in the middle of an insect infestation. It is not infrequent in these cases for cockroaches to be the insect around which the delusions center, and it is not uncommon for that delusion to be caused by a previous, actual encounter with roaches. An entomologist with the Los Angeles Public Health Department named William Waldron wrote in 1962 that he had seen over one hundred such cases in only five years (Schrut, 1962). As an example he described a case history in which a family of four in the City of Angels had seen a few *Periplaneta americana* in their home. From there, they went on to develop the collective delusion that bedbugs and lice were also present, in addition to a great quantity of cockroaches. Eventually, the wife took the children and left the husband and the house, unable to bear living there.

While these kinds of delusions may be excessive manifestations of the stress that having a few cockroaches can cause, there is a certain anxiety, for most people, when they are forced to share their living space with *Blattarians*. It is hard to say just what it is about cockroaches that gives so many people that little frisson, that extra bit of aversion and repulsion that makes roaches even less tolerable than spiders, flies, mosquitoes, slugs, and all the other slimy, nasty, filthy inhabitants of the insect world, but there is undeniably *something*. Since the beginning of time, at least as the human race knows it, cockroaches have  accompanied us, and for most of that time it is probably safe to say they were not welcome. Yet, our relationship with them is one of mutual nurturing—we provide them garbage to eat, and they eat it. Without them, we would have even more to dump in our already overflowing waste disposal sites, and without us, they would be out in the back country scavenging for their daily grub just like almost all of the world's vast number of cockroach species, which never have contact with humans.

At the level of racial consciousness, one of the deterrents to the human race's perpetrating a nuclear holocaust has been the notion that if we finally wiped ourselves out, or mutated the race beyond recognition with radioactive fallout, the cockroach would survive us, would inherit what remained, and would, somehow, manage to thrive on it, finally and fully appropriating the world that we consider rightfully ours. It is one thing to know that you can wipe out all humanity with a push of a button, eliminate the world as we know it, but another thing entirely when you realize that such aggression on behalf of great greed, or even a grand abstract political or religious idea, would not be the end of everything, but only the world as *we* know it. If the whole planet is wiped out for a cause, reduced to just one more cold and lifeless rock circling the sun, out of service to some obscure nationalist principle, then it's an heroic gesture, but what's the point of just finishing off *homo sapiens* if the world, itself, will keep on turning, inhabited by cockroaches and other similarly hardy things like snapping turtles and catfish. Their presence keeps us on our toes, and provokes an innate response of fear and caution in us late-comer life-forms who are dependent on such a complex and limited environment for our survival. It would be the great cutting-off-our-noses-to-spite-our-faces to wipe ourselves out and leave the world to roaches, so the bomb remains undropped and while we continue despoiling the planet, we worry about it more and more. Perhaps this will not, finally, save even one of the unprecedented number of currently endangered species from vanishing, but the roach remains to remind us of a yet more adaptable species than our own. In short, when we've drowned in our own shit, roaches will be dining on it. When we've eliminated our own species from the planet, cockroaches will be here to enjoy the leftovers.

. . • •

Andy Warhol: I used to come home and I used to be so glad to find a little roach there to talk to I just... it was great to have... at least somebody was there to greet you at home, right? And then they just go away. They're great. I couldn't step on them. Do you step on them?
William Burroughs: Oh no—God, man! I either have a sprayer... Well, occasionally I get a water-bug in my place. There's something called Tat with a thin tube coming out from the nozzle and it makes this fine spray. If you see a waterbug you can just...

Warhol: You don't have any roaches in your new place?
Burroughs: Very rarely. I got rid of them all.

from a conversation at a dinner party in New York City, 1980, recorded by Victor Bockris in his book, *WITH WILLIAM BURROUGHS*

● ● ● ·

Just how much radiation can your average cockroach absorb before its ability to propagate the species is affected? No one is quite certain. Some researchers have come up with numbers like 9,600 rads (radiation absorbed dose) over 35 days, while humans die after 2 weeks of 1,000 rads. Cockroaches are reported to have survived the blasts at Hiroshima and Nagasaki, which provided about 1,200 rads to humans on the ground. What is certain is that they can keep on reproducing when hit with what would be enough to fry us, cause our hair to fall out in great clumps, and our teeth to crumble to dust.

Joseph Kunkel, a well-respected roach researcher, is an entomologist at the University of Massachusetts, where he maintains a bibliograpic web site with the latest papers about *Blattarians*. In an answer to my question about radioactivity, he wrote me:

> I have been told that cockroaches are more resistant to radiation but have not seen any publication that discusses it with any credibility. I can give only an opinion of my own. I have irradiated cockroaches and constructed killing curves for them using gamma irradiation. I have not compared their resistance to radiation with any other organism using the same equipment

and thus can not comment on any relative resistance based on hard data.

My opinion is that insects in general would be relatively resistant to radiation compared to noninsects, or nonarthropods more strictly. The lives of insects and other arthropods revolve around their molting cycles. During a molting cycle the cells of the insect divide usually only once. This is encoded in Dyar's Rule, i.e., insects double their weight at each molt and thus their cells need divide only once per molting cycle.

Now it just so happens that cells are most sensitive to radiation when they are dividing. That is the basis on which radiation is used to kill cancer cells. Cancer cells tend to divide more often than the other cells of our body. For a given dose of radiation you will kill more cancer cells than normal cells. With the right dose, with the right cancer, you can kill all the cancer cells while only killing some of the most rapidly dividing normal cells (i.e., bone marrow cells of our immune system and red blood cells generating tissue).

Now if a typical cockroach molts at most once a week, its cells usually divide within a forty-eight-hour period within that week. That means that about three-quarters of the cockroaches would not have cells that are particularly radiation sensitive at any one time. If a killing radiation is endured by a cockroach and human population, then three-quarters of the cockroaches might survive while none of the humans might survive since our blood stem-cells and immune stem-cells are dividing all the time.

> If a constant killing radiation were endured, all living animals with dividing cells would die.

David George Gordon echoes this last thought when he points out in his excellent book, THE COMPLEAT COCKROACH, that while a roach can survive a blast the strength of the bomb dropped on Hiroshima, it cannot withstand the much more powerful weapons of today. Nevertheless, it is not hard to imagine a scenario in which a few bombs are dropped here and there, destroying the fabric of our so-called civilization, while roaches carry on as usual, probably not even aware that anything of magnitude has occurred, no doubt mutating all the time so as to be able to better withstand some ambient radiation. Cockroaches have repeatedly demonstrated, if not flaunted, an enviable ability to perpetuate their species.

Take, for instance, the much ballyhooed Biosphere 2, an expensive experiment in which $200 million was spent between 1984 and 1991 to build a self-enclosed, self-sustaining ecosystem in Arizona, a miniature replica of earth that contained forests, streams, an ocean, soil, animals, and plants. Eight scientists were sealed in this virtual Noah's Ark in 1991. They were expected to stay two years and, in theory, were equipped to grow their own food, breathe air recirculated by plants, and drink water created and purified by natural processes. The idea was that if earth could be replicated in miniature, perhaps there would be hope for humans even if they despoiled the original biosphere, our planet, to the point where it was uninhabitable.

Eighteen months later, Biosphere 2 had to be opened so oxygen could be pumped in, otherwise there would not have been enough to support the humans inside. In less than a year and a half after it had been sealed, the oxygen concentration inside had fallen from 21 percent to 14 percent, the level that would be found, ordinarily, at an altitude of 17,500 feet. Biosphere 2 was resealed, but at the end of two years it was clear that the miniature world was far from self-sustaining. All the plant-pollinating insects had died, meaning there was no possibility of reproduction for those few species of plants that had survived. Of the twenty-five small animals that started the experiment, nineteen had gone extinct, but the original populations of ants, cockroaches, and katydids were thriving. How long they might have been able to carry on in Biospehere 2 is a matter for conjecture, but while the humans were in danger of perishing from oxygen deprivation and were having to work like fiends to keep vines and weeds from overrunning their carefully cultivated plots of soil, the roaches were doing just fine.

In their comment on the Biosphere experiment, Joel Cohen and David Tilman wrote in *SCIENCE:* "The major retrospective conclusion that can be drawn is simple. At present there is no demonstrated alternative to maintaining the viability of Earth. No one yet knows how to engineer systems that provide humans with life-supporting services that natural eco-systems produce for free. . . . Despite its mysteries and hazards, Earth remains the only known home that can sustain life."

By which, of course, they mean human life. To be sure, there are those who look to the stars, who hold that when we're through driving this planet to wrack and ruin, we shall have the capacity to carry ourselves to others and make them habitable. Perhaps. But one thing is for certain: cockroaches will come with us. They have shown themselves to be perfectly capable of surviving the hardship of liftoff, and proven that once in outer space they can develop perfectly well without gravity. In 1998, a small colony of American cockroaches was sent into space as an experiment, but even before that there were rumors of stowaways on space shots. A roach was spotted by prelaunch workers on Apollo XII and was never definitively killed or removed. The mission not only went into space, but sent a vehicle down to the surface of the moon. Whether there was a roach on board, whether in fact roaches have already colonized the moon, remains an open question. There has also been talk of roaches being sighted on the MIR space station.

Regardless of when cockroaches first actually shucked off the bonds of gravity, what is certain is that *Periplaneta americana* were along for the most celebrated of the 1990s' space flights, the Discovery mission that took 77-year-old John Glenn back into space for 134 orbits around the earth. The roaches were on board thanks to Carolyn Harden, a biology teacher at Duval High School in Lanham, Maryland, and her students. The question that the students wanted answered was whether cockroaches could withstand a space launch, and if so whether they could reproduce once they were in space. NASA agreed to the experiment and the American Institute of Aeronautics and Astronautics put up $5,000. There was another $25,000 or so of in-kind donations of equipment and labor, and it took the students a year to design and prepare the experiment. They took three adults, three juveniles, and three egg cases and put them into a canister. They kept three more of each at the school as a control population. The canister was equipped with dog biscuits, a water delivery system, a mirror, and a video camera that took pictures every two hours.

The whole thing had to be delivered to NASA three months before liftoff, which, of course, in *Periplaneta americana* time is about twenty-five or thirty human years. A lot happened while they were waiting. Looking at the videotape, it appears that two of the egg cases hatched out and that almost all the roaches, adults and instar nymphs, died before Discovery was launched, said Harden. "We think they died of thirst," she told me on the phone, shortly after the recovery of Discovery's payload. "We had a simple system rigged up using a cotton wick in a bottle of water, but something must have gone wrong with it."

That was the bad news. However, two of the roaches, both of which began the journey as nymphs, not only survived until the launch, but made the 134 orbits and returned in good shape. One was an adult when the canister was opened, and one was still an instar, but close to adulthood.

"The question was what would happen if they get aboard by accident, as is said to have happened on the MIR," Harden told me. "We proved that they can definitely survive the rigors of the launch. Both the survivors were in the nymph compartment and the one has already become an adult, a male. The other survivor is still a nymph and we can't tell its sex. If it molts into a female we'll mate them and carry on the population."

An experiment that will not, unfortunately, be carried out. On the very weekend that students, teacher, and roaches journeyed to the Goddard Space Flight Center to meet the astronauts and review their research with them, the nymph died. Nevertheless, the students have decided to take a female from the roaches that were born among the ground control group and mate it with the survivor of the space flight resulting in a hybrid group that will be compared with those descended from roaches that stayed on the ground. And, Harden holds out hopes for putting another group of *Periplaneta americana* on board a space flight with a more efficiently designed water delivery system.

.   •   •   ●

Alas, poor roach. Industry, courage, thought, philosophy—they are the gifts which come, night by night, in the kitchen sink, from the roach as he rises hand in hand with mankind on the long, long climb from savagery to civilization.

Henry H. Curran,
New York City deputy mayor
to Fiorello LaGuardia, 1938

Cockroaches help to clean up the garbage that humans leave around, and they have performed one other service for us—they have died by the millions to increase our body of scientific knowledge about life and how it manifests itself. And, they have done so without even the minimum safeguards of comfort and humane treatment that federal law requires of scientists who use dogs, cats, monkeys, and other animals in their laboratories. Cockroaches are not covered by the Laboratory Animal Welfare Act.

Much of the scientific literature, while couched in the rather dense terminology of the discipline, is chilling. Take the following description of an experiment designed to observe "...neural activity from the central nervous system of a suspended cockroach while the animal can still walk, groom, and perform other acts....The animal is prepared by being anesthetized with carbon dioxide. The wings are removed and the insect impaled on a pair of parallel No. 7 insect pins clamped in a holder....By pulling out and cutting away the alimentary canal, the fat body, and reproductive organs, nearly the entire ventral nerve cord can be exposed in the bottom of the body cavity. The animal is given a lightweight (hollowed out) Styrofoam ball to hold as it recovers from the anesthetic....Cockroaches seem remarkably little affected by the procedure just described. Preparations remain viable for many hours, even overnight....Evisceration seems to have little effect on the animal's movements" (Delcomyn, 1976).

The fact is, there are not a whole lot of other positive interactions between humans and cockroaches. Of course, there are not a lot of negative ones either, given the minimum amount of disease they pass along and food they rob or contaminate. For a long time it was believed that they ate bedbugs (*Cimus*

the mating ritual

lectularius L.), and could be used, themselves, as a biological deterrent. Bedbugs were, during the nineteenth century and first part of the twentieth, the number one household pest, far worse than roaches because they lived by biting people. Once glass and screens began to be common in windows and people were able to keep mosquitoes outside the house, the last domestic pest that actually attacked people in their own homes was the bedbug. A number of books for general readers about insects asserted that roaches ate bedbugs, and there were reports of cockroaches actually being brought ashore by sailors in various ports and given to locals for eliminating bedbug populations.

Nevertheless, after an extensive review of the literature, Roth and Willis concluded that while cockroaches may occasionally dine on a bedbug, there is no evidence to conclude that they would be of any use in controlling them. There was, for instance, an experiment in which *Blattella germanica* were starved and then placed in a container with bedbugs. They did not eat any of them. *Periplaneta americana* would eat the tender young bedbugs, but left the adults, with their fully developed exoskeletons, alone (Roth, 1957).

Technology may be about to open up other ways for cockroaches to serve humanity's needs, and Madagascar hissers are at the forefront of the research. Building on many years of escape-response research into which way cockroaches move when puffs of air are blown on their antennae or their cerci, Japanese researchers, supported by a $5 million grant from their government, strapped small backpacks on American cockroaches through which an electrical "stimulus" could be delivered. The idea was that if a roach's movements could be controlled, it might eventually be used to go into places where humans cannot, for instance to explore buildings destroyed by weather, or terrorist attack, and locate survivors (Holzer, 1997). A cockroach with a camera is a good idea for getting into such disaster zones, but only if the insect's movements can be directed.

Figure 19. Bluetooth-enabled RoboRoach from Backyard Brains.

Figure 20. A bearded dragon eating a Madagascar hissing cockroach in Wimberley, Texas. Photo © Shannon Marlow du Plessis.

Scientists at the University of Michigan have been building on the work of the Japanese bio-robot experiments, but they have substituted Madagascar hissing cockroaches for *Periplaneta americana* and found that they could steer 10 to 20 percent of them with electrical stimuli applied to either an antenna or a cercus (Moore, 1998). Simultaneously, a wristwatch size data package and voice communicator was developed that could, in theory, be strapped to a roach's back. The hissers can carry more weight than that totaled by both the electrical stimulator and the instrument package, says the lead scientist on the project, University of Michigan entomologist Thomas Moore, who loaded up his hissing cockroaches with quarters on their backs to see how much weight they could move.

Like his Japanese counterparts, Moore envisions using roaches to carry microcameras or voice transmitters to places too small for people to reach, the sorts of spaces left after hurricanes or tornadoes, but he prefers the hissers to the American cockroaches used in Japan. They are better suited to the task because they are stronger, slower, and have no wings that need to be removed or might interfere with technological apparatus. Even the cockroach's ability to withstand radiation could, eventually, be used in our favor. Imagine a disaster at a nuclear power plant. How better to assess the damage than to send in a Madagascar hisser with a miniature camera on its back?

The research still has a long way to go, however. Only three of fifty hissers tested by the team responded well enough to follow a zig-zag "track" some ten feet long, and no one understands why there was such a difference in performance between one individual and another—one of the lucky trio who performed well was featured in a report that was aired on CNN about the research. How long the cockroaches could be made to repeat their successful perambulations without excessive electrical stimulus is another question that remains to be answered. Nevertheless, the research group was encouraged enough that they plan to continue their work.

. . • ●

The long winter draws to a close.

The big, hundred-year-old sycamores that line the street outside my window are budding out, it's time to take overcoats and heavy jackets to the dry cleaners and mothball them until next year. The great good news is that a pair of the Madagascar hissing cockroaches, one male and one female, have survived into another spring, along with myself and all my loved ones. The two hissers seem to be picking up, recovering from their winter torpor. They have more appetite and even do some exploring at night, perhaps even doing what is necessary to propagate their community after I am in bed and not watching. The sad news is that the world is overrun with civil and border wars, and people go on fighting and killing each other at a horrible rate, as far as ever from stopping the great wheel of suffering from its seemingly endless turnings.

A cockroach that has not been your mother at some time in the past is difficult to find.

. . • ●

At Savatthi. There the Blessed One said: "From an inconstruable beginning comes transmigration. A beginning point is not evident, though beings hindered by ignorance and fettered by craving are transmigrating & wandering on. A being who has not been your mother at one time in the past is not easy to find....A being who has not been your father...your brother...your sister...your son...your daughter at one time in the past is not easy to find.

"Why is that? From an inconstruable beginning comes transmigration. A beginning point is not evident, though beings hindered by ignorance and fettered by craving are transmigrating & wandering on. Long have you thus experienced stress, experienced pain, experienced loss, swelling the cemeteries—enough to become disenchanted with all fabricated things, enough to become dispassionate, enough to be released."

Mata Sutta
Samyutta Nikaya, XV 14-19
translated from the Pali by Thanissaro Bhikkhu

bibliography

Alali, F.Q., et al. "Annonaceous Acetogenins as Natural Pesticides: Potent Toxicity against Insecticide-Susceptible and -Resistant German Cockroaches." *Journal of Economic Entomology* 91 (1998): 641-649.

Algren, Nelson. *The Man with the Golden Arm.* New York: Seven Stories Press, 1997.

Altner, H., et al. "Relationship Between Structure and Function of Antennal Chemo-, Hygro-, and Thermoreceptive Sensilla in *Periplaneta americana.*" *Cell and Tissue Research* 176 (1977): 389-405.

Angier, Natalie. "How Can You Like a Lowly Roach?" *The New York Times,* 12 March 1991, sec. C1.

Anonymous. "Activists Seek Restrictions on Dursban." *Chemical Market Reporter,* 25 November 1996, p. 6.

Anonymous. "Mother of Baby Who Choked to Death on Cockroach Files Suit." *Jet,* 2 June 1997, 2.

Appel, Arthur G. "Laboratory and Field Performance of Consumer Bait Products for German Cockroach Control." *Journal of Economic Entomology* 83 (1990): 153-159.

_____. "Daily Pattern Trap Catch of German Cockroaches in Kitchens." *Journal of Economic Entomology* 91 (1998): 1136-1141.

Atxaga, Bernardo. *Obabakoak.* New York: Pantheon, 1992.

Ball, Harold J. "Photosensitivity in the Terminal Ganglion of *Periplaneta americana* (L.)." *Journal of Insect Physiology* 11 (1965): 1311-1315.

_____. "The Receptor Site for Photic Entrainment of Circadian Activity Rhythms in the Cockroach *Periplaneta americana.*" *Annals of the Entomological Society of America* 64 (1971): 1010-1015.

Barcay, S.J., et al. "Influence of Insecticide Treatment on German Cockroach Movement and Dispersal within Apartments." *Journal of Economic Entomology* 83 (1990): 142-147.

Barker, Will. *Familiar Insects of America.* New York: Harper & Bros., 1960.

Barth, Robert H. Jr. "Hormonal Control of Sex Attractant Production in the Cuban Cockroach." *Science* 133 (1960): 1598-1599.

Barth, Robert H., and Paul Sroka. "Initiation and Regulation of Oöcyte Growth by the Brain and Corpora Allata of the Cockroach, *Nauphoeta cinerea.*" *Journal of Insect Physiology* 21 (1975): 321-330.

Beebe, William. "Migration of Insects (Other than Lepidoptera) through Portachuelo Pass, Rancho Grande, North-Central Venezuela." *Zoologica* 36 (1951): 255-266.

Bell, W.J., and G. R. Sams. "Aggressiveness in the Cockroach *Periplaneta americana.*" *Behavioural Biology* 9 (1973): 581-593.

Bell, W.J., and R.E. Gorton. "Informational Analysis of Agonistic Behaviour and Dominance Hierarchy Formation in a Cockroach, *Nauphoeta cinera.*" *Behaviour* 67 (1978): 217-235.

Bell, William J., and K.G. Adiyodi, eds. *The American Cockroach.* New York: Chapman and Hall, 1981.

Belpré, Pura. *Perez y Martina.* New York: Viking Penguin, 1991.

Belt, Thomas. *The Naturalist in Nicaragua.* London: Edward Bumpus, 1888.

Berenbaum, May. *Bugs in the System.* Reading, Mass.: Addison Wesley, 1995.

Bernton, H.S., and Halla Brown. "Cockroach Allergy II: The Relation of Infestation to Sensitization." *Southern Medical Journal* 60 (1967): 852-855.

Berthold, Robert Jr., and Billy Ray Wilson. "Resting Behavior of the German Cockroach, *Blattella germanica*." *Annals of the Entomological Society of America* 60 (1997): 347-351.

Bockris, Victor. *With William Burroughs*. London: Fourth Estate, 1996.

Brady, John. "Control of the Circadian Rhythm of Activity in the Cockroach." *Journal of Experimental Biology* 47 (1967): 153-178.

Breed, Michael D., and Christine D. Rasmussen. "Behavioural Strategies During Intermale Agonistic Interactions in a Cockroach." *Animal Behaviour* 28 (1980): 1063-1069.

Bret, Brian L., et al. "Influence of Adult Females on Within-Shelter Distribution Patterns of *Blattella germanica*." *Annals of the Entomological Society of America* 76 (1983): 847-852.

Campbell, Frank L., and June D. Priestley. "Flagellar Annuli of *Blatella germanica*—Changes in Their Numbers and Dimensions During Postembryonic Development." *Annals of the Entomological Society of America* 63 (1970): 81-88.

Catesby, Mark. *The Natural History of Carolina, Florida and the Bahama Islands*. London: Royal Society of London, 1747.

Center for Public Integrity. *Unreasonable Risk: The Politics of Pesticides*. Washington, D.C.: Center for Public Integrity, 1998.

Cochran, Donald G. "Monitoring for Insecticide Resistance in Field-Collected Strains of the German Cockroach." *Journal of Economic Entomology* 82 (1989): 336-341.

———. "Toxic Effects of Boric Acid on the German Cockroach." *Experientia* 51 (1995): 561-563.

Cohen, Joel E., and David Tilman. "Biosphere 2 and Biodiversity: The Lessons So Far." *Science* 274 (1996): 1150-1151.

Company, Flavia. *Ni Tu, Ni Jo, Ni Ningú*. Barcelona: Ediciones 62, 1998.

Cook, Benjamin J., and Gerald G. Holt. "Neurophysiological Changes Associated with Paralysis Arising from Body Stress in the Cockroach, *Periplaneta americana*." *Journal of Insect Physiology* 20 (1974): 21-40.

Corbett, J.R. *The Biochemical Mode of Action of Pesticides*. New York: Academic Press, 1974.

Cornwell, P.B. *The Cockroach, Vol. 1.* London: Hutchinson & Co., 1968.

Cowan, Frank. *Curious Facts in the History of Insects.* Philadelphia: J.B. Lippincott & Co., 1865.

Cunliffe, F. "Biology of Cockroach Parasite, *Pimeliaphilus pdoapolipophagus* Trägardh, with a discussion of the genera *Pimeliaphilus* and *Hirstiella.*" *Proceedings of the Entomological Society of Washington* 54 (1952): 153-169.

Dangsheng, Liang, et al. "Field and Laboratory Evaluation of Female Sex Pheromone for Detection, Monitoring, and Management of Brownbanded Cockroaches (Dictyoptera: Blattellidae)." *Journal of Economic Entomology* 91 (1998): 480-485.

Delcomyn, Fred. "An Approach to the Study of Neural Activity During Behavior in Insects." *Journal of Insect Physiology* 22 (1976): 1223-1227.

Denic, Nebojsa, et al. "Cockroach: The Omnivorous Scavenger." *The American Journal of Forensic Medicine and Pathology* 18 (1997): 177-180.

Dong, K.E. "A Single Amino Acid Change in the Para Sodium Channel Protein is Associated with Knockdown-Resistance (kdr) to Pyrethroid Insecticides in German Cockroach." *Insect Biochemistry and Molecular Biology* 27 (1997): 93-100.

Dreisig, Hans, and Erik Tetens Nielsen. "Circadian Rhythm of Locomotion and Its Temperature Dependence in *Blatella germanica.*" *Journal of Experimental Biology* 54 (1971): 187-198.

Dunn, Ashley. "Wary of the Dangers of Killing Insecticides, Scientists Have Developed Techniques for Killing the Ultimate Urban Pest in Safer, Ecologically Sensitive Ways." *The New York Times*, 24 April 1994, sec. 14, p. 1.

Ebbett, Raymond H., and Donald Cochran. "Inheritance of Resistance to Pyrethroids in the German Cockroach." *Journal of Economic Entomology* 90 (1997): 1458-1461.

Ebeling, W., et al. "The Influence of Repellancy on the Efficacy of Blatticides. 1. Learned Modification of Behavior of the German Cockroach." *Journal of Economic Entomology* 56 (1966): 1374-1387.

_____. "The Influence of Repellancy on the Efficacy of Blatticides. III. Field Experiments with German Cockroaches with Notes on Three Other Species." *Journal of Economic Entomology* 61 (1968): 751-761.

Ebeling, W., and D.A. Reirson. "Effect of Population Density on Exploratory Activity and Mortality Rate of German Cockroaches in Choice Boxes." *Journal of Economic Entomology* 63 (1970): 351-355.

Edelson, Edward. "On the Trail of a Roach Remedy." *Newsday*, 8 December 1998, sec. C3.

Eisner, T. "Spray Mechanism of the Cockroach *Diploptera punctata*." *Science* 128 (1938): 147-149.

Evans, Howard Ensign. *Life on a Little-known Planet*. New York: E.P. Dutton and Co., 1968.

Fleet, R.R., et al. "Studies on the Population Ecology of the Smokybrown Cockroach, *Periplaneta fuliginosa*, in a Texas Outdoor Environment." *Environmental Entomology* 7 (1978): 807-814.

French, Vernon, and John Domican. "The Regeneration of Supernumerary Cockroach Antennae." *Journal of Embryology and Experimental Morphology* 67 (1982): 153-165.

Frishman, Austin M. *The Cockroach Combat Manual*. New York: William Morrow and Co., 1980.

Frishman, Austin M., and Edward Alcamo. "Domestic Cockroaches and Human Bacterial Disease." *Pest Control*, June 1997.

Garcia, Daniel P., et al. "Cockroach Allergy in Kentucky: A Comparison of Inner City, Suburban, and Rural Small Town Populations." *Annals of Allergy* 72 (1994): 203-208.

Gates, Mary Frances, and W.C. Allee. "Conditioned Behavior of Isolated and Grouped Cockroaches on a Simple Maze." *Journal of Comparative Psychology* 15 (1933): 331-358.

Gordon, David George. *The Compleat Cockroach*. Berkeley: Ten Speed Press, 1996.

Griffiths, J.T., and O.E. Tauber. "Fecundity, Longevity and Parthenogenesis of the American Roach, *Periplaneta americana*." Referred to in Louis M. Roth and Edwin Willis, *The Reproduction of Cockroaches*. (Washington, D.C.: Smithsonian Institution, 1954), 11.

Guenther, Konrad. *A Naturalist in Brazil*. London: George Allen & Unwin, 1939.

Gulati, A.N. "Do Cockroaches Eat Bed Bugs?" *Nature*, 13 June 1930, 858.

Guthrie, D.M., and A.R. Tindall. *The Biology of the Cockroach*. New York: St. Martin's Press, 1968.

Harington, Donald. *The Cockroaches of Stay More*. New York: Harcourt Brace Jovanovich, 1989.

Harker, J.E. "Control of Diurnal Rhythms of Activity in *Periplaneta americana* L." *Nature* 175 (1955): 733.

Hassell, Greg. "Combat Contest Crawls in Creepiest." *The Houston Chronicle*, 18 June 1997, sec. C1.

Heslop, J.P., and J.W. Ray. "The Reaction of the Cockroach *Periplaneta americana* L. to Bodily Stress and DDT." *Journal of Insect Physiology* 3 (1959): 395-401.

Holzer, R., and I. Shimoyama. "Locomotion Control of a Bio-robotic System Via Electric Stimulation." *Proceedings of the IEEE/RSJ International Conference on Intelligent Robots and Systems* (September 1997): 1514-1519.

Horridge, G.A. "Learning of Leg Position by the Ventral Nerve Cord in Headless Insects." *Proceedings of the Royal Society of London, Series B* 157 (1962): 33-52.

Ishii, Shoziro. "Aggregation of the German Cockroach, *Blattella germanica*." *Control of Insect Behavior By Natural Products*, eds. Wood, et al. (New York: Academic Press, 1970).

Janssen, Werner A., and Stanley E. Wedberg. "The Common House Roach, *Blattella germanica* Linn., as a Potential Vector of *Salmonella Thyimurium* and *Salmonella Typhosa*." *American Journal of Tropical Medicine and Hygiene* 1 (1952): 337-343.

Kaakeh, Walid, and Gary W. Bennett. "Evaluation of Trapping and Vacuuming Compared with Low-impact Insecticide Tactics for Managing German Cockroaches in Residences." *Journal of Economic Entomology* 90 (1997): 976-982.

Kaakeh, Walid, et al. "Comparative Contact Activity and Residual Life of Juvenile Hormone Analogs Used for German Cockroach (Dictyoptera: Blattellidae) Control." *Journal of Economic Entomology* 90 (1997): 1247-1253.

Kaliner, Michael A. "Asthma Deaths: A Social or Medical Problem?" *Journal of the American Medical Association* 269 (1993): 1994-1995.

Kalm, Peter. *Peter Kalm's Travels in North America*. New York: Dover, 1964.

Kattan, Meyer, et al. "Characteristics of Inner-city Children with Asthma: The National Cooperative Inner-city Asthma Study." *Pediatric Pulmonology* 24 (1997): 253-262.

Kellert, Stephen R. "Human-Animal Interactions: A Review of American Attitudes to Wild and Domestic Animals in the Twentieth Century." Quoted in *Animals and People Sharing the World*, ed. Andrew N. Rowan (Hanover, N.H.: University Press of New England, 1988), 137-175.

Kevan, D. Keith McE. "The Terrestrial Arthropods of the Bermudas: An Historical Review of Our Knowledge." *Archives of Natural History* 10 (1981): 1-29.

Kevler, Bettyann. *Females of the Species*. Cambridge: Harvard University Press, 1986.

Koehler, Philip G., et al. "German Cockroach Infestations in Low-income Apartments." *Journal of Economic Entomology* 80 (1987): 446-450.

Kohl; James Vaughn, and Robert T. Francouer. *The Scent of Eros: Mysteries of Odor in Human Sexuality*. New York: Continuum, 1995.

Laing, Frederick. *The Cockroach: Its Life-history and How To Deal with It*. London: The British Museum, 1930.

Larsen, G.S., et al. "Effects of Load Inversion in Cockroach Walking." *Journal of Comparative Physiology* 176 (1995): 229-238.

Lawson, Fred A. "Structural Features of the Oothecae of Certain Species of Cockroaches." *Annals of the Entomological Society of America* 44 (1951): 269-285.

Lehrer, Samuel B., et al. "Comparison of Cockroach Allergenic Activity in Whole Body and Fecal Extracts." *Journal of Allergy and Clinical Immunology* 87 (1991): 574-580.

Linden, Christopher H., et al. "Acute Ingestions of Boric Acid." *Clinical Toxicology* 24 (1986): 269-279.

Lipton, G.R., and D.J. Sutherland. "Feeding Rhythms in the American Cockroach, *Periplaneta americana*." *Journal of Insect Physiology* 16 (1970): 1757-1767.

Lispector, Clarice. *The passion according to G.H.* Minneapolis: University of Minnesota Press, 1997.

Longo, Nicholas. "Probability-Learning and Habit-Reversal in the Cockroach." *American Journal of Psychology* 77 (1964): 29-41.

Lovell, Kathryn L., and E.M. Eisenstein. "Dark Avoidance Learning and Memory Disruption by Carbon Dioxide in Cockroaches." *Physiology and Behavior* 10 (1973): 835-840.

Marlatt, C.L. *Cockroaches*. Washington, D.C.: U.S. Department of Agriculture, circular 51, 2d series, 1902.

Marquis, Don. *the lives and times of archy & mehitabel*. New York: Doubleday and Co., 1950.

McCarthy, Cormac. *The Orchard Keeper*. New York: Vintage Books, 1993.

McKittrick, F.A. "Evolutionary Studies of Cockroaches." *Cornell University Agricultural Experiment Station Memorandum* 389, 1964.

Mechling, Jay. "From archy to Archy, Why Cockroaches Are Good to Think." *Southern Folklore* 48 (1991): 121-140.

Metzger, Roland. "Behavior" in *Understanding and Controlling the German Cockroach*, edited by Michael Rust, et al. New York: Oxford University Press, 1995, 49-76.

Miall, L.C., and Alfred Denny. *The Cockroach*. London: Lovell Reeve & Co., 1886.

Moore, Patricia J. "Odour Conveys Status on Cockroaches." *Nature* 389, 4 September 1997, 25.

Moore, Thomas E., et al. "Directed Locomotion in Cockroaches: `Biobots´." *Acta Entomologia Slovenica* 6 (1998): 71-78.

Mosely, Henry Nottidge. *Notes by a Naturalist: An Account of Observations Made during the Voyage of H.M.S. Challenger Round the World in the Years 1872-1876*. London: MacMillan, 1879.

Mrozek, Slawomir. *El Árbol*. Barcelona: Quaderns Crema, 1998.

Mullins, Donald E., and Donald G. Cochran. "Nitrogenous Excretory Materials from the American Cockroach." *Journal of Insect Physiology* 19 (1973): 1007-1018.

Mullins, D.E., and C.B. Keil. "Paternal Investment of Urates in Cockroaches." *Nature* 283 (1980): 567-569.

Negus, Tracy F., and Mary H. Ross. "The Response of German Cockroaches to Toxic Baits: Strain Differences and the Effects of Selection Pressure." *Entomologia Experimentalis et Applicata* 82 (1997): 247-253.

Nichols, Arthur. "The Cockroach." *Nature* 3 (1870): 107-108.

Nishikawa, Michiko, et al. "Central Projections of the Antennal Cold Receptor Neurons and Hygroreceptor Neurons of the Cockroach *Periplaneta americana." Journal of Comparative Neurology* 361 (1995): 165-176.

Nishino, Chikao, et al. "Electroantennogram Responses of the American Cockroach to Germacrene D Sex Pheromone Mimic." *Journal of Insect Physiology* 23 (1977): 415-419.

Norris, D. M. "A Molecular and Submolecular Mechanism of Insect Perception of Certain Chemical Information in Their Environment." *Internal Collection of the French National Center for Scientific Research* 265 (1976): 81-102.

Owens, J.M., and G.W. Bennett. "German Cockroach Movement within and between Urban Apartments." *Journal of Economic Entomology* 75 (1982): 570-573.

Patton, R.L., et al. "The Excretory Efficiency of the American Cockroach, *Periplaneta Americana* L." *Journal of Insect Physiology* 3 (1959): 251-256.

Perry, Donald. *Life Above the Jungle Floor.* New York: Simon and Schuster, 1986.

Pettit, L.C. "The Effect of Isolation on Growth in the Cockroach (*Blattella germanica*) L." *Entomology News* 51 (1940): 293.

Plateau, Felix. *Note Sur Les Phenomènes de la Digestion Chez la Blatte Américaine.* Brussels: L'Académie Royale de Belgique, 1876.

Pollart, S.M., et al. "Identification, Quantitation and Purification of Cockroach Allergens Using Monoclonal Antibodies." *Journal of Allergy and Clinical Immunology* 87 (1991): 511-521.

Potera, Carol. "Working the Bugs Out of Asthma." *Environmental Health Perspectives* 105 (1997): 1192-1194.

Pritchatt, Derrick. "Avoidance of Electric Shock by the Cockroach *Periplaneta americana." Animal Behavior* 16 (1968): 178-185.

Puckett, Newbell Niles. *Folk Beliefs of the Southern Negro.* Chapel Hill: University of North Carolina Press, 1926.

Qadri, M.A.H. "The Life-history and Growth of the Cockroach *Blatta orientalis,* Linn." *Bulletin of Entomological Research* 29 (1938): 263-276.

Rau, P. "Food Preferences of the Cockroach, *Blatta orientalis,* Linn." *Entomology News* 56 (1945): 276-278.

Reagan, Douglas P., and Robert B. Waide, eds. *The Food Web of a Tropical Rain Forest.* Chicago: University of Chicago Press, 1996.

Rehn, James, A.G. "Man's Uninvited Fellow Traveler—The Cockroach." *Science Monthly* 61 (1945): 265-276.

Robinson, W.H. *Urban Entomology.* London: Chapman and Hall, 1996.

Romano, Jay. "Approaches to Killing Roaches." *The New York Times,* 3 August 1997, sec. 9, p. 3.

Rosenstreich, David L., et al. "The Role of Cockroach Allergy and Exposure to Cockroach Allergen in Causing Morbidity among Inner-city Children with Asthma." *The New England Journal of Medicine* 336 (1997): 1356-1363.

Ross, Mary H. "Genetic Study of Nonavoidance of a Pyrethroid Residue by German Cockroaches." *Journal of Economic Entomology* 90 (1997): 1243-1246.

_____. "Response of Behaviorally Resistant German Cockroaches to the Active Ingredient in a Commercial Bait." *Journal of Economic Entomology* 91 (1998): 150-151.

Ross, Mary H., et al. "Population Growth and Behaviour of *Blatella germanica* in Experimentally Established Shipboard Infestations." *Annals of the Entomological Society of America* 77 (1984): 740-752.

Ross, Mary H., and Brian L. Bret. "Effects of Propoxur Treatment on Populations Containing Susceptible and Resistant German Cockroaches." *Journal of Economic Entomology* 79 (1986): 338-349.

Ross, Mary H., and Donald E. Mullins. "Biology." In *Understanding and Controlling the German Cockroach,* edited by Michael Rust, et. al. New York: Oxford University Press, 1995, 21-47.

Roth, Louis M. "The Evolution of Male Tergal Glands in the Blattaria." *Annals of the Entomological Society of America* 62 (1969): 176-208.

_____. "Interspecific Mating in Blattaria." *Annals of the Entomological Society of America* 63 (1970): 1282-1285.

Roth, Louis M., and Edwin Willis. *The Reproduction of Cockroaches.* Washington, D.C.: Smithsonian Institution, 1954.

_____. *The Medical and Veterinary Importance of Cockroaches.* Washington, D.C.: The Smithsonian Institution, 1957.

_____. *The Biotic Associations of Cockroaches.* Washington D.C.: The Smithsonian Institution, 1960.

Roth, Louis M. and George Dateo Jr. "Uric Acid in the Reproductive System of Males of the Cockroach *Blattella germanica*." *Science* 146 (1964): 782-784.

Ruderman, Wendy. "Bugs Don't Lie." *The New York Times*, 1 June 1997, sec. 13, p. 3.

Rust, Michael K., and Donald Reierson. "Using Pheromone Extract to Reduce Repellancy of Blatticides." *Journal of Economic Entomology* 70 (1977): 34-38.

Rust, Michael K., et al., eds. *Understanding and Controlling the German Cockroach*. New York: Oxford University Press, 1995.

Sakuma, M., and H. Fukami. "The Aggregation Pheromone of the German Cockroach, *Blatella germanica* (L.): Isolation and Identification of the Attractant Components of the Pheromone." *Applied Entomology and Zoology* 25 (1990): 355-368.

Sarpong, Sampson B., and Theodore Karrison. "Season of Birth and Cockroach Allergen Sensitization in Children with Asthma." *Journal of Allergy and Clinical Immunology* 101 (1998): 566-568.

Schafer, R. "Postembryonic Development in the Antenna of the Cockroach, *Leucophaea maderae*: Growth, Regeneration and the Development of the Adult Pattern of Sense Organs." *Journal of Experimental Zoology* 183 (1973): 353-364.

Schafer, Rollie, and Thomas V. Sanchez. "Antennal Sensory System of the Cockroach, *Periplaneta americana*: Development and Morphology of Sense Organs." *Journal of Comparative Neurology* 149 (1973): 335-354.

Schal, C. "*Blaberus gigabteus* (cucaracha, Giant Cockroach, Giant Drummer, Cockroach of the Divine Face) and *Xestoblatta hamata* (Cucaracha)." *Costa Rican Natural History*, ed. Daniel H. Janzen. Chicago: University of Chicago Press, 1983.

Schal, C., and A.S. Chiang. "Hormonal Control of Sexual Receptivity in Cockroaches." *Experientia* 51 (1995): 994-998.

Scharrer, Berta. "The Relationship between Corpora Allata and Reproductive Organs in Adult *Leucophaea Maderae*." *Endocrinology* 38 (1946): 46-55.

Schrader, Ann. "CU Sets Up Test Battle of the Bugs." *Denver Post*, 22 July 1998, sec. A1.

Schrut, Albert H., and William G. Waldron. "Psychiatric and Entomological Aspects of Delusory Parasitosis." *Journal of the American Medical Association* 186 (1963): 429-430.

Scudder, Samuel Hubbard. "Revision of the American Fossil Cockroaches." *Bulletin of the United States Geological Survey* 124 (1895).

Siebert, Charles. "The Lives They Lived: Bertha Scharrer, What the Roaches Told Her." *The New York Times Magazine*, 31 December 1995, p. 26.

Siegel, Sheldon C. "History of Asthma Deaths from Antiquity." *Journal of Allergy and Clinical Immunology* 80 (1987): 458-462.

Silverman, J., et al. "Hydramethylnon Uptake by *Blattella germanica* by Coprophagy." *Journal of Economic Entomology* 84 (1991): 176-180.

Silverman, Jules, and Donald Bieman. "Glucose Aversion in the German Cockroach, *Blattella germanica*." *Journal of Insect Physiology* 39 (1993): 925-933.

Sims, Michael. *Darwin's Orchestra*. New York: Henry Holt & Co., 1997.

Smith, Gavin, ed. *Sayles on Sayles*. New York: Faber & Faber, 1998.

Smith, L.M. II, et al. "Insecticides for Cockroach Management." *Journal of Economic Entomology* 90 (1997): 1232-1242.

_____. "Comparison of Conventional and Targeted Insecticide Application for Control of Smokybrown Cockroaches in Three Urban Areas of Alabama." *Journal of Economic Entomology* 91 (1998): 473-479.

Stecklow, Steve. "New Food-Quality Act Has Pesticide Makers Doing Human Testing." *The Wall Street Journal*, 28 September 1998, p.1.

Stek, Michael, et al. "Retention of Bacteria in the Alimentary Tract of the Cockroach, *Blattella germanica*." *Journal of Environmental Health* 41 (1979): 212-213.

Stern, Michael, et al. "Regeneration of Cercal Filiform Hair Sensory Neurons in the First-Instar Cockroach Restores Escape Behaviour." *Journal of Neurobiology* 33 (1997): 439-458.

Stix, Gary. "Roach Wars." *Scientific American* 271 (1994): 85.

Suto, C., and N. Kumada. "Secretion of Dispersion-inducing Substance by the German Cockroach, *Blattella germanica* (L.)." *Applied Entomological Zoology* 16 (1981): 120-133.

Szymanski, J.S. "Modification of the Innate Behavior of Cockroaches." *Animal Behavior* 2 (1912): 81-90.

Tarshis, I. Barry. "The Cockroach—A New Suspect in the Spread of Infectious Hepatitis." *American Journal of Tropical Medicine and Hygiene* 11 (1962): 705-711.

Turner, C.H. "An Experimental Investigation of an Apparent Reversal of the Responses to Light of the Roach (*Periplaneta orientalis* L.)." *Biology Bulletin* 23 (1912): 371-386.

Wassmer, Gary T., and Terry L. Page. "Photoperiodic Time Measurement and a Graded Response in a Cockroach." *Journal of Biological Rhythms* 8 (1993): 47-56.

Wassmer, Gary T., et al. "Photoperiodic Regulation of Hemolymph Protein in the Woodroach *Parcoblatta pennsylvanica*." *Journal of Insect Physiology* 42 (1996): 851-858.

Wharton, D.R.A., et al. "Blood Volume and Water Content of the Male American Cockroach, *Periplaneta americana*, L.—Methods and the Influence of Age and Starvation." *Journal of Insect Physiology* 11 (1965): 391-404.

Wharton, Martha L., and D.R.A. Wharton. "The Production of Sex Attractant Substance and of Öothecae by the Normal and Irradiated American Cockroach, *Periplaneta americana*, L." *Journal of Insect Physiology* 1 (1957): 229-239.

Wilyeto, E.P., and G.M. Bousch. "Attraction of the German Cockroach, *Blattella germanica*, to Some Volatile Food Components." *Journal of Economic Entomology* 76 (1983): 752-756.

Yamada, Minoru. "Electrophysical Investigation of Insect Olfaction." *Control of Insect Behavior By Natural Products*, eds. Wood, et al. New York: Academic Press, 1970.

Yamamoto, Robert. "Collection of the Sex Attractant from Female American Cockroaches." *Journal of Economic Entomology* 56 (1963): 119-120.

Ye, Shuping, and Christopher M. Comer. "Correspondence of Escape-Turning Behavior with Activity of Descending Mechanosensory Interneurons in the Cockroach, *Periplaneta americana*." *Journal of Neuroscience* 16 (1996): 5844-5853.

index

albino, 40
as allergen, 108-11
antennae of, 31-34, 55-56, 71
attitudes toward, 42-43, 62, 157-58
Biosphere 2, survival in, 162-63
blood of, 5-6, 15, 16, 17
brains of, 36, 93 fig. 17, 94, 99
breathing of, 15-16
cleanliness of, 29-31
contact, preference for, 95
description of, 6, 8
digestive system of, 88, 89 fig. 14
disease, as transmitters of, 18, 112-15, 116-17
eyes of, 33
fat body of, 6, 13-14, 14 fig. 5, 15
flying abilities of, 11-12
in folklore, 120-22, 123
as food, 4, 83-84, 88, 90
food of, 6-7, 84-88, 89 fig. 15
fossils of, 7, 7 fig. 1, 66
geography and habitat of, 8
head of, 33 fig. 7
hormones of, 37, 39, 40-41, 53
humans, as threats to, 115-16, 116 fig. 18
identifying, 19

medical uses of, 119-20
metabolic processes of, 15
molting of, 37-39, 38 fig. 8, 60, 96
motion detectors of, 35-36
movement of, 96-99
nervous system of, 36
observe, how to, 63
order of, 8
origins of, 7-8
as pets, 42
pheromones of, 53, 54, 55, 67-71, 98
phylum of, 37-38
in popular culture, 123-24
radiation on, effect of, 160-62, 169
in scientific experiments, 166
sleeping habits of, 90-91, 91 fig. 16
on space flights, 163-65
species, number of, 8, 19
study of, 9
technology, helping humans through, 167-69, 168 fig. 19
ventral nerve of, 35-36, 37, 41-42
water, importance of, 16-18
See also cockroaches, species of; cockroach, history and use of the word; pest control; reproduction processes, of cockroaches; travelers,

disease, as transmitters of, 112, 113, 114
eating habits of, 6, 87-88, 167
molting of, 38, 39, 66-67
pest control, 133, 139-40, 148
in popular culture, 124
reproduction processes of, 54, 55, 56, 61, 66-67
traveling habits of, 72-73, 76
Gordon, David George, 162
Gromphadorhina portentosa.
See Madagascar hissing (*Gromphadorhina portentosa*) cockroaches

Harden, Carolyn, 164-65
Harington, Donald, 31
Harker, Janet, 92
Hartnack, H., 85
Hearn, Lafcadio, 119
hemolymph, 15, 16, 17, 36, 37
Honegger, Willi, 40-41, 42

Insect growth regulators (IGR), 147
insecticides. *See* pest control
instars, defined, 39
Integrated pest management (IPM), 146-47
Ishii, Shoziro, 67

Journal of Allergy and Clinical

Immunology, 110
Journal of Experimental Biology, 92
Journal of Insect Physiology, 17
Journal of Neuroscience, 34

Kafka, Franz, 100-101
Kalm, Peter, 134
Kunkel, Joseph, 160

Last Carousel, The (Algren), 5
lateral spiracles, 15-16
Ledoux, A., 67
Les Bizarreries Des Races Humaines (Coupin), 83
Leucophaea maderae, 36, 37, 66
Linnaeus, Carolus, 10
Lispector, Clarice, 82-83
Lives and Times of archy & mehitabel, The (Marquis), 99

Madagascar hissing *(Gromphadorhina portentosa)* cockroaches, 20-21, 20 fig. 6, 167-69, 168 fig. 20
Magazine of Natural History, 74
Man With The Golden Arm, The (Algren), 5
Marlatt, C.L., 86, 87
Marquis, Don, 99-100
Matarese, Marlene, 39-40
Medical and Veterinary Importance of Cockroaches, The (Roth and Willis), 18, 112
Melching, Jay, 120-21